Contents

There once lived a man
who learned how to slay dragons
and gave all he possessed
to mastering the art.

After three years
he was fully prepared but,
alas, he found no opportunity
to practise his skills.

Dschuang Dsi.

As a result he began
to teach how to slay dragons.

René Thom.

London Mathematical Society Lecture Note Series. 17

Differentiable Germs and Catastrophes

TH. BRÖCKER & L. LANDER

Universität Regensburg
Fachbereich Mathematik

Cambridge University Press
Cambridge
London · New York · Melbourne

Published by the Syndics of the Cambridge University Press
The Pitt Building, Trumpington Street, Cambridge CB2 1RP
Bentley House, 200 Euston Road, London NW1 2DB
32 East 57th Street, New York, N.Y. 10022, USA
296 Beaconsfield Parade, Middle Park, Melbourne 3206, Australia

Library of Congress Catalogue Card Number: 74-17000

ISBN: 0 521 20681 2

First published 1975
Reprinted 1976

Printed in Great Britain
at the University Printing House, Cambridge
(Euan Phillips, University Printer)

Foreword

In the summer semester of 1972 I gave a course of lectures on the local theory of differentiable maps at the University of Freiburg. These lectures have formed the basis for the first thirteen chapters of the book, the next three chapters having been written for a summer school organised by the Studienstiftung des deutschen Volkes. My students were responsible for removing many mistakes from the original manuscript which has now been translated into English by L. Lander. He has also made a number of improvements and corrections and provided the last chapter together with its pictures and list of publications. The later chapters discuss a subject which has been the real motivation for writing the book: classical catastrophe theory.

We have both profited greatly from a lecture course on catastrophe theory by K. Jänich, given in Regensburg during the winter semester 1971/72, which contained most of the information and pictures presented in chapter 17.

A small number of copies of the German text of the present book were printed for our students under the title: Der Regensburger Trichter, Band 3, Differenzierbare Abbildungen.

On the pages that follow, the reader will not find any new results or methods. Our purpose is to make it easier for those students, who have properly understood the basic lecture courses on analysis and possess a basic knowledge of algebra, to learn about recent work on differentiable maps, in particular, the mysteries of catastrophe theory.

What are the following pages about?

Let $f : \mathbf{R}^n \to \mathbf{R}^k$ be a differentiable map. What can be said in general about $f^{-1}\{0\}$, that is, about the solution set of a system of non-linear equations? To start with one refers to a theorem of Whitney and Sard's theorem, given in 2.1 and 3.3, in particular one discovers that interesting structure can only be found for 'generic' sets of maps.

Of special interest are the stable differentiable maps, where f is called stable if for a 'small perturbation' $\delta : \mathbf{R}^n \to \mathbf{R}^k$ there are invertible transformations such that the diagram

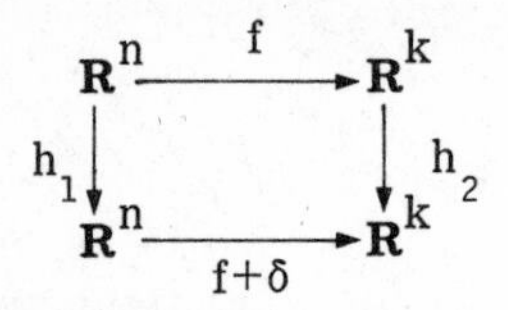

commutes. In fact, one expects that natural forms must be described by stable maps because everything in nature is subject to small disturbances. Is 'almost every' map stable? How is the concept of stability to be interpreted?

Any introduction to analysis explains that a differentiable germ $f : (\mathbf{R}, 0) \to (\mathbf{R}, 0)$ with non-vanishing Taylor expansion at the origin can be transformed into the first non-vanishing term by a suitable coordinate change. In higher dimensions, when is a germ determined by a finite part of its Taylor expansion (up to equivalence under coordinate transformations)?

Those are a few of the questions which are discussed below. Perhaps the reader will thereby be encouraged to join in the task of clarifying and understanding some of the ideas of R. Thom.

Regensburg, Spring 1974 Theodor Bröcker

1·Germs with constant rank

Literature: J. Dieudonné: Foundations of modern analysis, Academic Press (1969).

1.1. Suppose $A \subset \mathbf{R}^n$ is an arbitrary subset. A map $f : A \to \mathbf{R}^k$ is called differentiable if there is an open set $U \subset \mathbf{R}^n$ and a map $F : U \to \mathbf{R}^k$, such that $A \subset U$ and $F|A = f$, and such that the partial derivatives of F of every order exist and are continuous.

In what follows, our main interest will be in the local properties of maps. To make this more precise, we need the following definition: suppose $x \in A \subset \mathbf{R}^n$, V is a set, and $\mathfrak{F}$ is the set of pairs (U, f), where U is open in $\mathbf{R}^n$, $x \in U$, and $f : A \cap U \to V$ is an (arbitrary, continuous, differentiable, analytic, ...) map. Consider the following equivalence relation on the elements of $\mathfrak{F}$: for (U_1, f_1) and (U_2, f_2) in $\mathfrak{F}$, $(U_1, f_1) \sim (U_2, f_2)$ if and only if there is an open set $U \subset \mathbf{R}^n$, with $x \in U$ and $U \subset U_1 \cap U_2$, such that $f_1|U = f_2|U$. An equivalence class of this relation is called an (arbitrary, continuous, differentiable, analytic, ...) germ $\tilde{f} : (A, x) \to V$ at x (the tilde will frequently be omitted). Thus one speaks of germs of differentiable or analytic maps; further, since all subsets are defined by maps: $\mathbf{R}^n \to \{0, 1\}$, one may consider germs of subsets of $\mathbf{R}^n$. Germs behave much the same as maps, in particular, germs $\tilde{f}$, $\tilde{g}$ may be composed to give $\tilde{g} \circ \tilde{f}$:

$$(\mathbf{R}^n, x) \xrightarrow{\tilde{f}} (\mathbf{R}^m, y) \xrightarrow{\tilde{g}} \mathbf{R}^k$$

$$y = f(x)$$

$$\tilde{g} \circ \tilde{f} : (\mathbf{R}^n, x) \to \mathbf{R}^k .$$

If $f : U \to \mathbf{R}^m$, $g : V \to \mathbf{R}^k$ are representatives of $\tilde{f}$, $\tilde{g}$ then $f|f^{-1}(V) : f^{-1}(V) \to \mathbf{R}^m$ is a representative of $\tilde{f}$. The usual map-composite $g \circ f$ is defined on $f^{-1}(V) \subset U$ and this is a representative of

$\tilde{g} \circ \tilde{f}$. A differentiable germ $\tilde{f}: (\mathbf{R}^n, x) \to \mathbf{R}^k$ has a Jacobi-matrix $Df(x) : \mathbf{R}^n \to \mathbf{R}^k$ (a linear map). The germ $\tilde{f}$ has an inverse germ (with respect to '$\circ$') if and only if a representative f of $\tilde{f}$ has a local inverse in a sufficiently small neighbourhood of x. And this is the case if and only if $Df(x)$ is non-singular:

1.2. **Inverse-function theorem** (see Dieudonné). A germ $\tilde{f}: (\mathbf{R}^n, x) \to (\mathbf{R}^n, y)$ possesses an inverse germ $\tilde{f}^{-1} : (\mathbf{R}^n, y) \to (\mathbf{R}^n, x)$ if and only if the Jacobi-matrix $Df(x)$ is non-singular.

If $f : U \to \mathbf{R}^k$ is differentiable, $U \subset \mathbf{R}^n$, then the map $Df : U \to \mathbf{R}^{kn} = \{(k \times n)\text{ - matrices}\}$, $x \mapsto Df(x)$ is differentiable. The rank of f at x is defined to be the rank of $Df(x)$ and denoted $Rk_x f$. If $Rk_x f \geq s$, then a certain $(s \times s)$ - sub-matrix of $Df(x)$ has non-vanishing determinant. This determinant will be non-zero on a neighbourhood of x because Df and determinant are continuous. Hence the rank of f is never smaller than s on a neighbourhood of x, the rank of f cannot fall locally, and so the map $U \to \mathbf{Z}$, $x \mapsto Rk_x f$ is lower semicontinuous. Thus for any germ $\tilde{f}: (\mathbf{R}^n, x) \to \mathbf{R}^k$, there is a corresponding lower semi-continuous germ $Rkf : (\mathbf{R}^n, x) \to \mathbf{Z}$, $y \mapsto Rk_y f$.

One important consequence which we shall deduce from the inverse-function theorem is the following:

1.3. **The rank theorem** (see Dieudonné). Let $\tilde{f}: (\mathbf{R}^n, x) \to (\mathbf{R}^m, y)$ be a germ with constant rank (this means that the germ Rkf is the germ of a constant map) then there exist invertible germs $\tilde{\phi} : (\mathbf{R}^n, x) \to (\mathbf{R}^n, 0)$ and $\tilde{\psi} : (\mathbf{R}^m, y) \to (\mathbf{R}^m, 0)$, such that the germ

$$\tilde{\psi} \circ \tilde{f} \circ \tilde{\phi}^{-1} : (\mathbf{R}^n, 0) \to (\mathbf{R}^m, 0)$$

is represented by the map $(x_1, \ldots, x_n) \mapsto (x_1, \ldots, x_k, 0, \ldots, 0)$ where $k = Rk_x f$.

Forgetting about germs, this result says that if a map $f : U_1 \to \mathbf{R}^m$, defined on the neighbourhood U_1 of x, has constant rank on a possibly smaller neighbourhood U_2 of x, then on a still smaller neighbourhood U_3 of x the map f has the given form with respect to suitable coordinates on $\mathbf{R}^n$ and $\mathbf{R}^m$.

Proof. Without loss of generality $x = y = 0$. Suppose f is a representative of $\tilde{f}$, with constant rank k. There will be a $(k \times k)$-submatrix of Df which is regular at the origin. By change of coordinates, that is, by applying local diffeomorphisms (invertible differentiable maps) the submatrix

$$(\partial f_i/\partial x_j), \quad 1 \leq i, \ j \leq k$$

may be assumed regular at $0 \in \mathbf{R}^n$, and hence regular on a neighbourhood of the origin.

Define the germ $\tilde{\phi} : (\mathbf{R}^n, 0) \to (\mathbf{R}^n, 0)$ by

$$(x_1, \ldots, x_n) \mapsto (f_1(x), \ldots, f_k(x), x_{k+1}, \ldots, x_n)$$

where f has components $(f_1, \ldots, f_m)$.

$$D\phi = \left[\begin{array}{c|c} \multicolumn{2}{c}{\partial f_i/\partial x_j} \\ \hline 0 & \begin{matrix} 1 & & 0 \\ & \ddots & \\ 0 & & 1 \end{matrix} \end{array}\right] \begin{matrix} \}\, k \\ \\ \}\, n-k \end{matrix}$$

with column blocks of width k and $n-k$.

$$\det(D\phi) = \det(\partial f_i/\partial x_j)_{1 \leq i,\, j \leq k}$$

$$\neq 0.$$

Hence $\tilde{\phi}$ is an invertible germ and the diagram

$$\begin{array}{ccc} (\mathbf{R}^n, 0) & \xrightarrow{f} & (\mathbf{R}^m, 0) \\ & \phi \searrow \quad \nearrow & \\ & (\mathbf{R}^n, 0) & \end{array} \qquad g = f \circ \phi^{-1} \qquad \begin{array}{ccc} (x_1, \ldots, x_n) & \xrightarrow{f} & (f_1, \ldots, f_m) \\ \searrow & & \nearrow \\ (f_1, \ldots, f_k, & x_{k+1}, \ldots, x_n) & \\ & \| & \\ (z_1, \ldots & & , z_n) \end{array}$$

shows that the germ $\tilde{g} = \tilde{f} \circ \tilde{\phi}^{-1}$ is represented by

$$z = (z_1, \ldots, z_n) \underset{g}{\mapsto} (z_1, \ldots, z_k, g_{k+1}(z), \ldots, g_m(z)).$$

The Jacobian matrix of g has the form

$$Dg = \left[\begin{array}{ccc|c} 1 & & 0 & \\ & \ddots & & 0 \\ 0 & & 1 & \\ \hline & ? & & A(z) \end{array}\right],$$

$$A(z) = (\partial g_j/\partial z_i) \quad k+1 \le j \le m, \; k+1 \le i \le n$$

and because $Rk(g) = Rk(Dg) = k$ in a neighbourhood of the origin, the matrix $A(z)$ must vanish on this neighbourhood. Hence, without loss of generality:

(*) $\quad \partial g_j/\partial z_i = 0$ for $k+1 \le j \le m, \; k+1 \le i \le n.$

Now apply a local transformation in the range $\mathbf{R}^m$, namely the germ $\tilde{\psi} : (\mathbf{R}^m, 0) \to (\mathbf{R}^m, 0)$ given by

$$\begin{bmatrix} y_1 \\ \vdots \\ y_k \\ y_{k+1} \\ \vdots \\ y_m \end{bmatrix} \mapsto \begin{bmatrix} y_1 \\ \vdots \\ y_k \\ y_{k+1} - g_{k+1}(y_1, \ldots, y_k, 0, \ldots, 0) \\ \vdots \\ y_m - g_m(y_1, \ldots, y_k, 0, \ldots, 0) \end{bmatrix}$$

The Jacobian of $\tilde{\psi}$ has the form

$$D\tilde{\psi} = \left[\begin{array}{ccc|ccc} \multicolumn{3}{c|}{\overbrace{\qquad\qquad}^{k}} & \multicolumn{3}{c}{\overbrace{\qquad\qquad}^{m-k}} \\ 1 & & 0 & & 0 & \\ & \ddots & & & & \\ 0 & & 1 & & & \\ \hline & & & 1 & & 0 \\ & ? & & & \ddots & \\ & & & 0 & & 1 \end{array}\right]$$

since $g_{k+1}(y_1, \ldots, y_k, 0, \ldots, 0)$ etc., do not depend on $y_{k+1}, \ldots, y_m$. Hence $\tilde{\psi}$ is invertible, and $\tilde{\psi} \circ \tilde{g}$ is represented by the composite

$$\begin{bmatrix} z_1 \\ \vdots \\ z_k \\ z_{k+1} \\ \vdots \\ z_n \end{bmatrix} \mapsto \begin{bmatrix} z_1 \\ \vdots \\ z_k \\ g_{k+1}(z) \\ \vdots \\ g_m(z) \end{bmatrix} \mapsto \begin{bmatrix} z_1 \\ \vdots \\ z_k \\ g_{k+1}(z) - g_{k+1}(z_1, \ldots, z_k, 0, \ldots, 0) \\ \vdots \\ g_m(z) - g_m(z_1, \ldots, z_k, 0, \ldots, 0) \end{bmatrix}.$$

Because of (*) the last $m - k$ components $g_{k+j}(z_1, \ldots, z_n) - g_{k+j}(z_1, \ldots, z_k, 0, \ldots, 0)$ of this composite vanish on an n-cube $|z_j| < \varepsilon$. Hence $\tilde{\psi} \circ \tilde{g}$ is represented by

$$(z_1, \ldots, z_n) \mapsto (z_1, \ldots, z_k, 0, \ldots, 0). \quad \checkmark$$

1.4. Application. Let $U \subset \mathbf{R}^n$ be an open subset, then $f: U \to \mathbf{R}^k$ is called a submersion if $\mathrm{Rg}_x f = k$ and an immersion if $\mathrm{Rg}_x f = n$ for all $x \in U$. By the rank theorem a submersion (immersion) has the form

$$(x_1, \ldots, x_n) \mapsto (x_1, \ldots, x_k)$$

$$((x_1, \ldots, x_n) \mapsto (x_1, \ldots, x_n, 0, \ldots, 0))$$

with respect to suitable coordinates. For, its rank cannot become any larger and is thus constant.

1.5. Definition. A subset $M \subset \mathbf{R}^n$ is called a differentiable submanifold of $\mathbf{R}^n$ of dimension $m \leq n$, if to each $x \in M$ there corresponds an invertible germ $\tilde{\phi} : (\mathbf{R}^n, x) \to (\mathbf{R}^n, 0)$, such that $\tilde{\phi}(M, x) = (\mathbf{R}^m, x) \subset (\mathbf{R}^n, x)$ ($\mathbf{R}^m$ linearly imbedded in $\mathbf{R}^n$ for $m \leq n$)

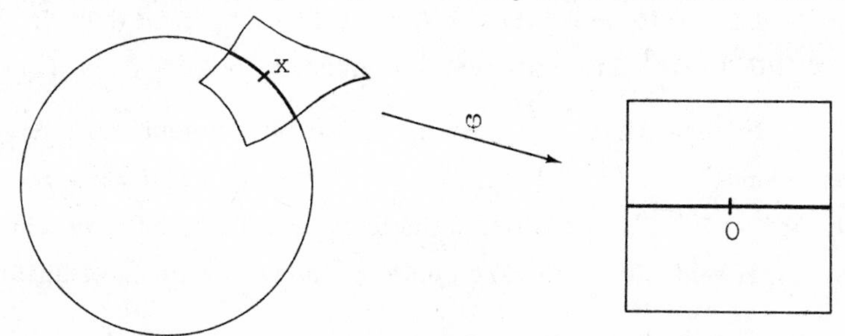

1.6. Example. $S^n = \{x \in \mathbf{R}^{n+1} | \langle x, x \rangle = 1\}$ is a differentiable submanifold of $\mathbf{R}^{n+1}$. The proof is left as an exercise.

1.7. Exercise. The set $LA(n, m; k) \subset LA(n, m) = \mathbf{R}^{nm}$ of $(m \times n)$-matrices with rank k in the space of all $(m \times n)$-matrices is a differentiable submanifold. Determine its dimension.

1.8. Definition. Let $f : \mathbf{R}^n \to \mathbf{R}^m$ be differentiable. A point $y \in \mathbf{R}^m$ is called a <u>regular value</u> of f if at each $x \in \mathbf{R}^n$, such that $f(x) = y$, the rank of f is m, that is, $Rk_x f = m$. Any value of f which is not regular is said to be a <u>critical value.</u> If $y \notin f(\mathbf{R}^n)$, then this definition makes y a regular value of f.

1.9. Theorem. <u>If y is a regular value of $f : \mathbf{R}^n \to \mathbf{R}^m$, then $f^{-1}\{y\} \subset \mathbf{R}^n$ is a differentiable submanifold of dimension $m - n$ (or empty).</u>

Proof. Let $x \in f^{-1}\{y\}$, so that $f(x) = y$ and $Rk_x f = n$. This means that the rank of f is locally constant at x. By the rank theorem, there are local differentiable transformations $\phi : (\mathbf{R}^n, x) \to (\mathbf{R}^n, 0)$ and $\psi : (\mathbf{R}^m, y) \to (\mathbf{R}^m, 0)$ such that the germ $\tilde{\psi} \circ \tilde{f} \circ \tilde{\phi}^{-1} = \tilde{f}_1$ has the form

$$f_1(x_1, \ldots, x_m, \ldots, x_n) = (x_1, \ldots, x_m).$$

The germ $\tilde{f}_1^{-1}\{0\} = \tilde{\phi}\tilde{f}^{-1}\tilde{\psi}^{-1}\{0\} = \tilde{\phi}\tilde{f}^{-1}\{y\}$ is the germ of the set $\{(0, \ldots, 0, x_{m+1}, \ldots, x_n)\}$ at the origin. ✓

1.10. Exercises.

1. Suppose that A is a symmetric $(n \times n)$-matrix and $0 \neq b \in \mathbf{R}$, prove that the set $M = \{x \in \mathbf{R}^n | x^t A x = b\}$ is either an $(n-1)$-dimensional submanifold of $\mathbf{R}^n$ or empty.

2. Let $f : \mathbf{R}^n \to \mathbf{R}^n$ be a differentiable map such that $f \circ f = f$. Prove that $f(\mathbf{R}^n) \subset \mathbf{R}^n$ is a differentiable submanifold.

3. In general, $f^{-1}\{y\}$ is not necessarily a submanifold - give a counter example.

If $M^m \subset \mathbf{R}^{m+k}$ is a differentiable submanifold and if we choose an invertible representative for every germ $\tilde{\phi}$ in the definition of differentiable submanifold, then:

Each $x \in M$ has a neighbourhood U_λ on which a differentiable map $\phi_\lambda : U_\lambda \overset{\rightarrow}{\cong} U'_\lambda \subset \mathbf{R}^m$ is defined, where U'_λ is an open subset.

If one wants to define a differentiable manifold without using an imbedding in $\mathbf{R}^{m+k}$, then the following definition presents itself:

1.11. A <u>differentiable manifold</u> is a topological space M, together with an open cover $\{U_\lambda \mid \lambda \in \Lambda\}$ and homeomorphisms $U_\lambda \to U'_\lambda \subset \mathbf{R}^m$ (U'_λ open), with the following properties:

(I)

$$\begin{array}{ccc} U_\lambda & \cap & U_\mu \\ \phi_\lambda \downarrow & & \downarrow \phi_\mu \\ \mathbf{R}^m \supset U'_{\lambda\mu} & \overset{\phi_{\lambda\mu}}{\dashrightarrow} & U'_{\mu\lambda} \subset \mathbf{R}^m, \end{array}$$

if $U_\lambda \cap U_\mu \neq \emptyset$, the map $\phi_{\lambda\mu}$ exists to make the diagram commute and is differentiable. Since $\phi_{\lambda\mu}\phi_{\mu\lambda} = \mathrm{Id}$, the map $\phi_{\lambda\mu}$ is a diffeomorphism (invertible). Obviously $U'_{\lambda\mu} = \phi_\lambda(U_\lambda \cap U_\mu)$ etc.

(II) M is Hausdorff and has a countable basis.

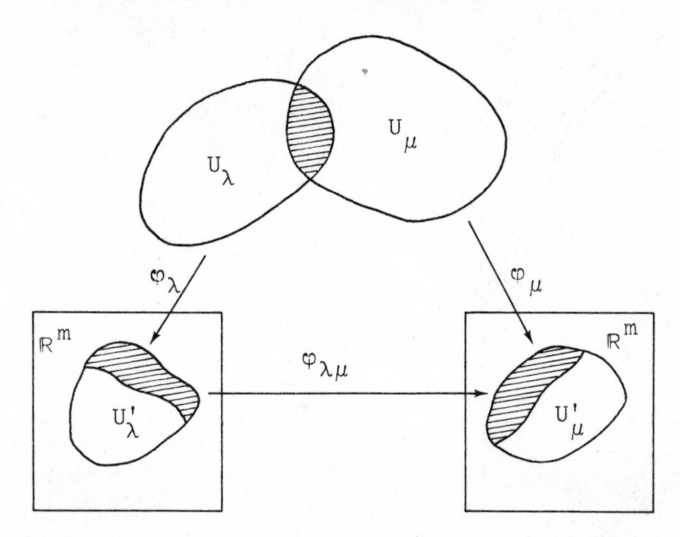

Many concepts are carried over from euclidean space to submanifolds of $\mathbf{R}^n$ (and to manifolds). For example, let $x \in M^m \subset \mathbf{R}^{m+k}$ then a function $f : M \to \mathbf{R}$ is <u>differentiable</u> at x if there is an invertible germ

$\tilde{\psi} : (\mathbf{R}^{m+k}, x) \to (\mathbf{R}^{m+k}, 0)$ such that

$$\tilde{\psi} : (M^m, x) \underset{\cong}{\to} (\mathbf{R}^m, 0)$$

and such that the germ $\tilde{f} \circ \tilde{\psi}^{-1}$ is differentiable at $0 \in \mathbf{R}^m \subset \mathbf{R}^{m+k}$.

In other words: M is covered by open sets U_λ, which may be identified with open subsets of $\mathbf{R}^m$ using coordinate transformations. A map defined on M is called differentiable (has rank r, etc.) if restricted to these open subsets - i.e. subsets of $\mathbf{R}^m$ up to transformation - it is differentiable (has rank r, etc.).

2·Regular values

Literature: J. Milnor: Topology from the differentiable viewpoint, Virginia (1969).

R. Narasimhan: Analysis on real and complex manifolds, Masson and Cie, Paris, and North-Holland, Amsterdam (1968).

S. Sternberg: Lectures on differential geometry, Prentice-Hall (1964).

The aim of this chapter is to prove the following theorem:

2.1. **Sard's theorem.** The Lebesgue-measure of the set of critical values of a differentiable map is zero.

A set with measure zero is defined below, but first of all consider an application: let $f : \mathbf{R}^n \to \mathbf{R}^m$ be differentiable. Then for almost all points $b \in \mathbf{R}^m$ (i. e. everywhere except for a set with Lebesgue measure zero) the following is true: $f^{-1}\{b\} \subset \mathbf{R}^n$ is a differentiable submanifold of dimension $n - m$. In other words:

For given $f = (f_1, \ldots, f_m)$ and for almost any choice of $b_i \in \mathbf{R}$, $1 \leq i \leq m$, the system of non-linear equations $f_i(x) = b_i$, $x \in \mathbf{R}^n$, $1 \leq i \leq m$, has an $(n-m)$-dimensional manifold for its solution set.

2.2. **Preliminaries.** Let $\{K_n\}_{n \in \mathbf{N}}$ be the set of balls in $\mathbf{R}^m$ with rational radius and rational coordinates at the centre (there are countably many!).

If $U \subset \mathbf{R}^m$ is open, then $U = \bigcup_{i \in T} K_i$ for a certain subset $T \subset \mathbf{N}$.

Proof. Let $x \in U$ and ε be small enough so that the ε-neighbourhood of x is contained in U. Choose K_i with centre at y for $|x - y| < \varepsilon/3$ and rational radius r, where $|x - y| < r < 2\varepsilon/3$. ✓

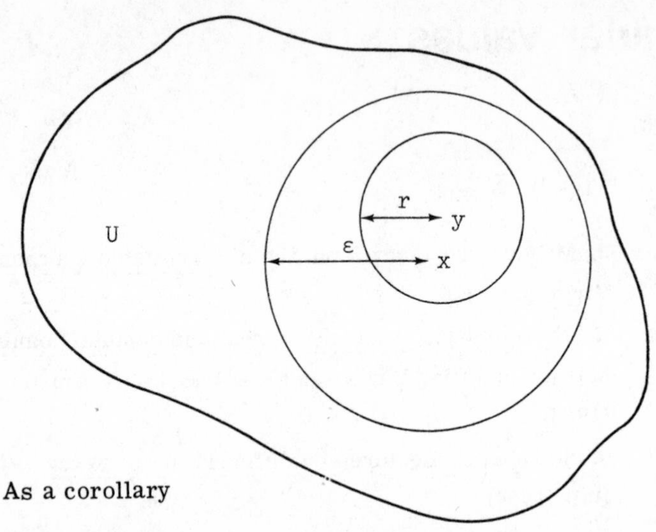

As a corollary

2.3. **Remark.** Let $X \subset \mathbf{R}^n$ be an arbitrary subset and $\{U_\lambda\}_{\lambda \in \Lambda}$ a family of open subsets of $\mathbf{R}^n$, such that $X \subset \bigcup_{\lambda \in \Lambda} U_\lambda$, then there is a countable subset $\Gamma \subset \Lambda$ such that $X \subset \bigcup_{\lambda \in \Gamma} U_\lambda$.

Proof. X is in the union of those K_n, which are contained in at least one U_λ, and there are only countably many such K_n. For each of these K_n choose a corresponding $U_{\lambda(n)}$ where $K_n \subset U_{\lambda(n)}$. The set X is contained in the union of the $U_{\lambda(n)}$. ✓

2.4. **Definition.** A subset $C \subset \mathbf{R}^n$ has measure zero, if for each $\varepsilon > 0$, there exists a sequence of cubes $W_i \subset \mathbf{R}^n$ such that

$$C \subset \bigcup_{i=1}^{\infty} W_i, \quad \text{and} \quad \sum_{i=1}^{\infty} |W_i| < \varepsilon.$$

Here $|W_i|$ is the volume of W_i, that is, $|W_i| = a^n$ where a is the side-length of W_i.

2.5. Clearly if $C = \bigcup_{\nu=1}^{\infty} C_\nu$ and each C_ν has measure zero, then C also has measure zero. For,

$$C_\nu \subset \bigcup_{i=1}^{\infty} W_i^\nu \quad \text{where} \quad \sum_{i=1}^{\infty} |W_i^\nu| < \frac{\varepsilon}{2^\nu}$$

and this implies

$$C \subset \bigcup_{i,\nu} W_i^\nu \quad \text{where} \quad \sum_{i,\nu} |W_i^\nu| < \sum_\nu \frac{\varepsilon}{2^\nu} = \varepsilon.$$

A similar argument shows that to define sets of measure zero, the W_i may be open or closed or they may be balls, cuboids, etc., instead of cubes.

2.6. **Lemma.** <u>If</u> $C \subset \mathbf{R}^n$ <u>has measure zero and</u> $f : C \to \mathbf{R}^n$ <u>is differentiable, then</u> $f(C)$ <u>has measure zero.</u>

Proof. Choose an open subset U with $C \subset U$ and a differentiable map $F : U \to \mathbf{R}^n$, with $F|C = f$. Since U is the union of a sequence of closed balls, it is no loss of generality to assume C is contained in a compact ball and that the covering of C by cubes is also contained in a (slightly larger) compact ball K, which is itself contained in U.

Now let

$$b = \max\left\{ \left| \frac{\partial F_i}{\partial x_j}(x) \right| \,\middle|\, x \in K \right\};$$

if the cube W has side-length a, then $|x_i - x_i^0| \le a$ for $x \in W$ implies

$$|F_i(x) - F_i(x^0)| \le b.n.a$$

so that $F(W)$ is contained in a cube of side-length $(b.n).a$. Therefore $|W| = a^n$ implies $F(W)$ lies inside a cube with volume

$$(bn)^n . a^n = (bn)^n . |W|$$

and $(bn)^n$ is independent of W. If $\Sigma |W_i| < \varepsilon/(bn)^n$ then the union of all the $F(W_i)$ is contained in a union of cubes with total volume smaller than ε. ✓

From what has been said it follows in particular that:

2.7. The property that a set $C \subset \mathbf{R}^n$ has measure zero is a <u>'local differential-topological'</u> property.

<u>Local</u> means that C has measure zero if and only if each point $x \in \mathbf{R}^n$ has a neighbourhood U such that $C \cap U$ has measure zero.
$\Rightarrow$: trivial, $U = \mathbf{R}^n$

$\Leftarrow$: Cover $\mathbf{R}^n$ with countably many such neighbourhoods U_n, then $C = C \cap (\bigcup_{n\in\mathbf{N}} U_n) = \bigcup_{n\in\mathbf{N}} (C \cap U_n)$ has measure zero.

Differential-topological means that diffeomorphisms transform sets of measure zero into sets of measure zero, and more generally, so do differentiable maps $\mathbf{R}^n \to \mathbf{R}^n$.

Warning. The property of having measure zero is not a topological property. There are homeomorphisms of the plane into the plane which map the unit (straight-line) interval onto a set with positive measure.

Proof? Exercise (see K. Mayrhofer: Inhalt und Mass, Springer 1952, and apply the Jordan curve theorem).

2.8. **Theorem** (Fubini). Let $\mathbf{R}_t^{n-1} = \{x \in \mathbf{R}^n | x_n = t\} \subset \mathbf{R}^n$, let $C \subset \mathbf{R}^n$ be compact and suppose $C \cap \mathbf{R}_t^{n-1}$ is thin in $\mathbf{R}_t^{n-1} \cong \mathbf{R}^{n-1}$ (i.e. has measure zero) for all $t \in \mathbf{R}$, then C is thin in $\mathbf{R}^n$.

Proof. We use the following

2.9. **Lemma.** An open covering of the interval $[0, 1]$ by subintervals contains a finite covering $[0, 1] \subset \bigcup_{j=1}^{k} I_j$ with $\Sigma |I_j| \leq 2$.

Proof of this lemma. Choose a finite, minimal subcover from the given open intervals, i.e. a subcover from which no interval may be omitted. Suppose this cover is $\{I_j\}$, $j = 1, \ldots, k$, where I_j has end points a_j, b_j. The numbering should be according to the size of the left-hand end point a_j. This is uniquely determined for if $a_i = a_j$, then either $b_i \leq b_j$ and (a_i, b_i) is redundant or $b_j < b_i$ and (a_j, b_j) is redundant (we exclude the trivial case $k \leq 2$). Further

$$a_i < a_{i+1} < b_i \leq a_{i+2}.$$

For, if the second inequality did not hold there would be a hole in the covering. As for the third inequality, observe $b_i < b_{i+1}$ (otherwise $(a_i, b_i) \supset (a_{i+1}, b_{i+1})$). Without this inequality one would have

$$(a_i, b_i) \cup (a_{i+2}, b_{i+2}) \supset (a_{i+1}, b_{i+1}).$$

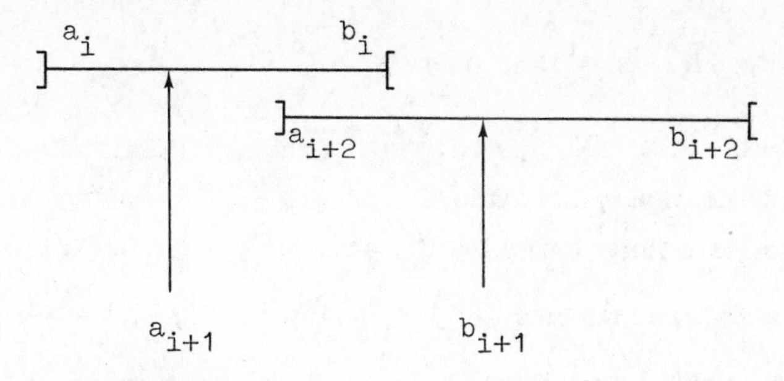

It now follows that

$$\begin{aligned}\Sigma(b_i - a_i) &= \Sigma(a_{i+1} - a_i) + \Sigma(b_i - a_{i+1})\\ &< \Sigma(a_{i+1} - a_i) + \Sigma(a_{i+2} - a_{i+1})\\ &< 2\end{aligned}$$

This proves the lemma. ✓

Now we come to Fubini's theorem:

Define C_t by $C_t \times \{t\} = C \cap (\mathbf{R}^{n-1} \times \{t\})$. Without loss assume $C \subset \mathbf{R}^{n-1} \times [0, 1]$, and C_t is thin for all $t \in [0, 1]$. Suppose $\varepsilon > 0$ is given. Find a cover of C_t by open cubes $\{W_t^i | i \in \mathbf{N}\}$ with total volume smaller than ε. Let $W_t = \bigcup_{i \in \mathbf{N}} W_t^i \subset \mathbf{R}^{n-1}$.

If x_n is the last coordinate then for fixed t the function $|x_n - t|$ is continuous on C and vanishes exactly on $C_t \times \{t\}$. Since $C - (W_t \times [0, 1])$ is compact, the function $|x_n - t|$, restricted to C, achieves a minimum value α.

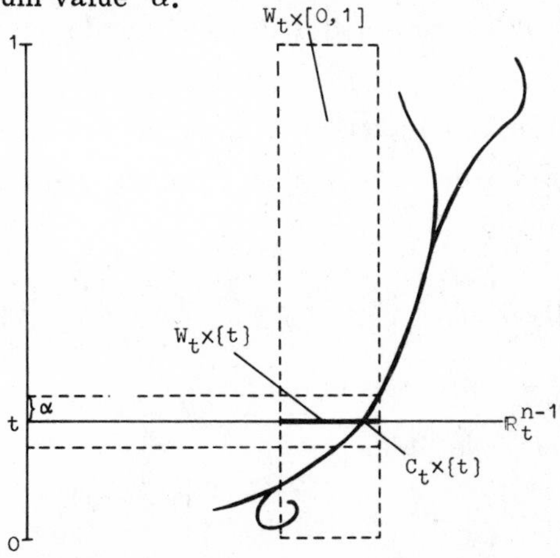

Therefore

$$C \cap \{x \in \mathbf{R}^n \mid |x_n - t| < \alpha\} \subset W_t \times I_t^\alpha$$

where $I_t^\alpha = (t - \alpha, t + \alpha)$.

The collection of intervals I_t^α covers $[0, 1]$, and so, from the lemma, there is a finite cover by I_t^α, say $\{I_j \mid j = 1, \ldots, k\}$, $I_j = I_{t_j}^\alpha$, with $\Sigma |I_j| \leq 2$. The cuboids $\{W_{t_j}^i \times I_j \mid j = 1, \ldots, k, i \in \mathbf{N}\}$ cover C and have total volume less than 2ε. ✓

2.10. **Extension of Fubini.** Instead of 'C compact' it is sufficient to assume C is a countable union of compact sets. Some examples of such sets are:

(I) closed sets: $C = \bigcup_{n \in \mathbf{N}} C \cap \{|x| \leq n\}$

(II) open sets: (Unions of countably many closed balls)

(III) Images of these sets under continuous maps: $\mathbf{R}^n \to \mathbf{R}^n$ (compact sets map to compact sets).

(IV) Countable unions of finite intersections of sets of these types.

Next we prove Sard's theorem:

2.11. <u>Let</u> $U \subset \mathbf{R}^n$ <u>be open,</u> $f : U \to \mathbf{R}^p$ <u>differentiable and let</u> $D = \{x \in U \mid \mathrm{Rk}_x f < p\}$ <u>be the set of critical points of</u> f, <u>then</u> $f(D)$ <u>is thin.</u>

Proof. Induction on n.

For $n = 0$, $\mathbf{R}^n = \{0\}$ and $f(U)$ is at most a point and the theorem is true.

For the inductive step let $D_i \subset U$ be the set of $x \in U$, where all partial derivatives of order $\leq i$ vanish. The D_i form a descending sequence of closed sets,

$$D \supset D_1 \supset D_2 \supset \ldots$$

We shall show that

(a) $f(D - D_1)$ is thin,

(b) $f(D_i - D_{i+1})$ is thin,

(c) $f(D_k)$ is thin for sufficiently large k.

Remark 1. All the sets to be considered fall into one of the four categories (I)-(IV) in the extension to Fubini's theorem.

Remark 2. For case (a) (and similarly for cases (b) and (c)) it is sufficient to prove that each point $x \in D - D_1$ has a neighbourhood V, such that $f(V \cap (D - D_1))$ is thin, since $D - D_1$ is covered by countably many such V.

Proof of (a). Assume $p \geq 2$, since for $p = 1$, we have $D = D_1$. Let $d \in D - D_1$. Since $d \notin D_1$, there is a partial derivative (without loss $\frac{\partial f_1}{\partial x_1}$) such that $\frac{\partial f_1}{\partial x_1}(d) \neq 0$.

Hence the map

$$h : U \to \mathbf{R}^n, \quad (x_1, \ldots, x_n) \mapsto (f_1(x), x_2, \ldots, x_n)$$

has Jacobian

$$D(h) = \left[\begin{array}{c|ccc} \frac{\partial f_1}{\partial x_1} & & \cdots & \\ \hline 0 & 1 & & 0 \\ \vdots & & \ddots & \\ 0 & 0 & & 1 \end{array}\right]$$

and $D(h)(d)$ is non-singular.

It follows that in a neighbourhood V of d, the map h is a coordinate transformation.

Consider the diagram

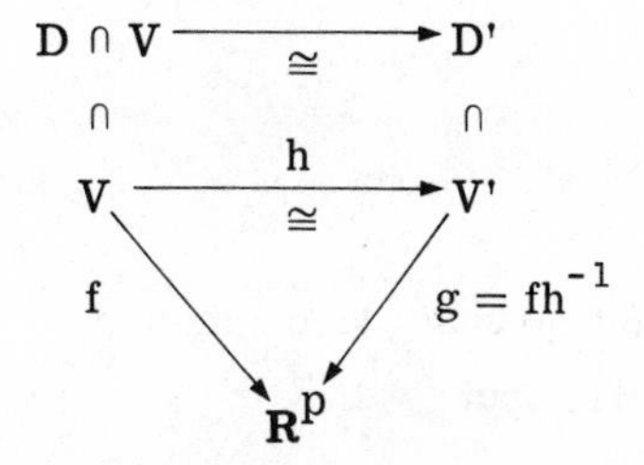

$$g(x_1, \ldots, x_n) = (x_1, g_2(x), \ldots, g_p(x)).$$

Obviously g maps the hyperplane $\{x_1 = t\}$ into the hyperplane $\{y_1 = t\}$. Let

$$g^t : (\{t\} \times \mathbf{R}^{n-1}) \cap V' \to \{t\} \times \mathbf{R}^{p-1}$$

be the restriction of g.

A point in $(\{t\} \times \mathbf{R}^{n-1}) \cap V'$ is a critical point for g exactly when it is critical for g^t, since the Jacobian matrix of g is

$$Dg = \left[\begin{array}{c|c} 1 & 0 \ldots 0 \\ \hline ? & \partial g_i^t / \partial x_j \end{array}\right] .$$

By induction, the set of critical values of g^t is thin in $\{t\} \times \mathbf{R}^{n-1}$, so that $g(D')$ has a thin intersection with each hyperplane $\{x | x_1 = t\} \subset \mathbf{R}^p$. By Fubini, $g(D')$ is thin.

Proof of (b). As in (a), if $d \in D_k - D_{k+1}$, there is a derivative of order $k + 1$ which does not vanish at d. Without loss

$$\frac{\partial^{k+1} f_1}{\partial x_1 \partial x_{s_2} \ldots \partial x_{s_{k+1}}} (d) \neq 0.$$

Let $w : U \to \mathbf{R}$ be the function

$$w(x) = \frac{\partial^k f_1}{\partial x_{s_2} \ldots \partial x_{s_{k+1}}} (x)$$

so that $w(d) = 0$ and $\frac{\partial w}{\partial x_1}(d) \neq 0$, since $d \in D_k$ and $d \notin D_{k+1}$. As before the map $h : U \to \mathbf{R}^n$

$$h(x) = (w(x), x_2, \ldots, x_n)$$

is a coordinate transformation on a neighbourhood V of d, and $h(D_k \cap V) \subset \{0\} \times \mathbf{R}^{n-1} \subset \mathbf{R}^n$. Once again, put

$$g = f \circ h^{-1} : V' = h(V) \to \mathbf{R}^p,$$

and

$$g^0 = g|\ldots : (\{0\} \times \mathbf{R}^{n-1}) \cap V' \to \mathbf{R}^p.$$

By induction, the set of critical values of g^0 is thin. But every point in $h(D_k \cap V)$ is critical for g^0 since all partial derivatives of g (and hence of g^0) of order $\leq k$ (in particular order 1) vanish. Hence $gh(D_k \cap V)$ is thin.

Proof of (c). Let $W \subset U$ be a cube with side-length a, and let $k > \frac{n}{p} - 1$, then we shall show that $f(W \cap D_k)$ is thin. Since U is a countable union of cubes, this is sufficient. Taylor's theorem gives

$$\text{(T)} \qquad \left.\begin{aligned} f(x+h) &= f(x) + R(x,\ h) \\ |R(x,\ h)| &\leq c.\ |h|^{k+1} \end{aligned}\right\} \quad \begin{aligned} &\text{for } x \in D_k \cap W \\ &x + h \in W, \end{aligned}$$

where c depends on W and f.

Split W into r^n cubes of side-length $\frac{a}{r}$. If W_1 is a cube in this decomposition, and W_1 contains $x \in D_k$, then each point in W_1 can be written as $x + h$ with

$$|h| \leq \frac{a\sqrt{n}}{r}.$$

Hence using (T), the set $f(W_1)$ is contained in a cube with side-length

$$2.c.\ (a\sqrt{n})^{k+1}/r^{k+1} = b/r^{k+1},$$

for a constant $b = b(f,\ W)$.

Altogether, these cubes have total volume $s \leq r^n .\ b^p/r^{p(k+1)}$, and for $p(k+1) > n$ this expression converges to zero as r increases. The total volume may therefore be made arbitrarily small by choice of a suitably fine decomposition. ✓

2.12. Exercises. 1. Extend Sard's theorem to maps between differentiable manifolds.

2. Let $M^m \subset \mathbf{R}^n$ be a differentiable submanifold and $L^{n-1} \subset \mathbf{R}^n$ a linear subspace. Then there is an affine subspace $A \subset \mathbf{R}^n$

which is parallel to L^{n-1} and such that $A \cap M^m$ is an $(m-1)$-dimensional submanifold.

2.13. Typical application. Let $D^n = \{x \mid |x| \leq 1\} \subset \mathbf{R}^n$. *Let* $f : D^n \to \mathbf{R}^n$ *be continuous, and* $f(S^{n-1}) \subset D^n$, *then* f *has a fixed point.* (Brouwer).

Sketch proof. Assume not, then $|x - f(x)| \geq \delta > 0$ for all x. Approximate f differentiably so that the approximation f_1 is fixed-point free (for example: $|f_1(x) - f(x)| < \delta/2$, for all x). The map $\phi : D^n \to S^{n-1}$, shown in the diagram,

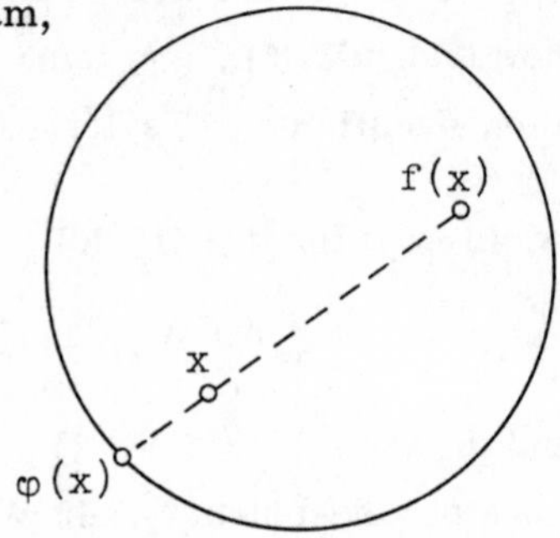

is differentiable, with $\phi | S^{n-1} = \text{id}$. Let $y \in S^{n-1}$ be a regular value of ϕ (Sard's theorem). Consider $M = \phi^{-1}\{y\} \subset D^n$. Its intersection $M \cap \mathring{D}^n$ with the interior of D^n is a differentiable submanifold with dimension 1. Using the rank theorem one finds coordinates on a neighbourhood of y such that f is given locally by

$$(y_1, \ldots, y_n) \mapsto (y_1, \ldots, y_{n-1}, 0)$$

and, also locally, S^{n-1} is the set $\{y_n = 0\}$. Hence y has a neighbourhood in M which is isomorphic to $[0, 1)$

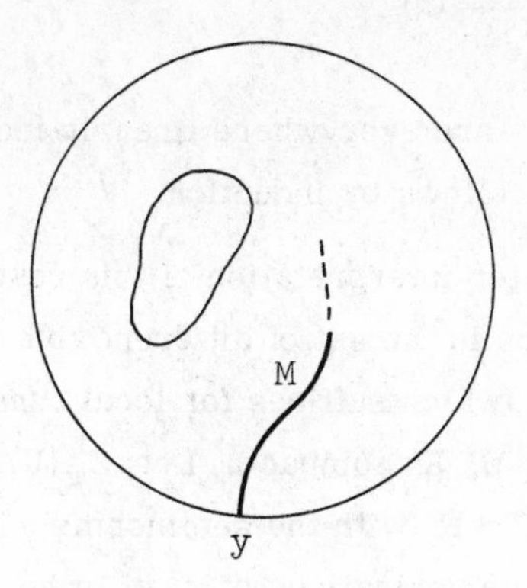

M is compact, hence it is a compact 1-dimensional manifold with boundary. Such manifolds are sums of circles and closed intervals (up to differentiable transformations). However M has only the one boundary point y. Contradiction. ✓

2.14. **A second application.** Let $f : \mathbf{R}^n \to \mathbf{R}^p$ be differentiable, $p \geq 2n$, then there is a linear map $A : \mathbf{R}^n \to \mathbf{R}^p$ with arbitrarily small norm, such that $f + A : \mathbf{R}^n \to \mathbf{R}^p$ is an immersion.

Proof. f is an immersion if the vectors $\{(\partial f/\partial x_i(x)) \,|\, i=1, \ldots, n\}$ are linearly independent at each point $x \in \mathbf{R}^n$.

Let $\{\partial f/\partial x_i \,|\, i=1, \ldots, s\}$ be linearly independent everywhere, $s \leq n$.

Consider

$$\phi : \mathbf{R}^s \times \mathbf{R}^n \to \mathbf{R}^p$$

$$\phi(\lambda_1, \ldots, \lambda_s, x) = \sum_{j=1}^{s} \lambda_j \frac{\partial f}{\partial x_j}(x) - \frac{\partial f}{\partial x_{s+1}}(x)$$

Since $s + n < p$, $\phi(\mathbf{R}^s \times \mathbf{R}^n)$ is thin. Let $a \in \mathbf{R}^p$, a small, $a \notin \phi(\mathbf{R}^s \times \mathbf{R}^n)$. Set $g(x) = f(x) + ax_{s+1}$. Then

$$\frac{\partial g}{\partial x_i} = \frac{\partial f}{\partial x_i} \quad \text{for } i \leq s$$

$$\frac{\partial g}{\partial x_{s+1}} = \frac{\partial f}{\partial x_{s+1}} + a$$

and for no $x \in \mathbf{R}^n$ do we find:

$$\sum_{i=1}^{s} \lambda_j \frac{\partial g}{\partial x_j}(x) = \frac{\partial g}{\partial x_{s+1}}(x) .$$

Hence the $\partial g/\partial x_i$ are everywhere linearly independent for $i \leq s + 1$. The theorem follows by induction. √

We can give a better interpretation of this result if we introduce the idea of neighbourhoods in the set of differentiable maps. Consider the following description (which suffices for local considerations):

Let $U \subset \mathbf{R}^n$, $K \subset U$, K compact. Let $C_K^k(U)$ be the set of differentiable maps $f : U \to \mathbf{R}$ with the seminorm: $|f|_K^k$ = the maximum value over K of any of the derivatives of f of order $\leq k$.

An ε-neighbourhood of f consists of all $g \in C_K^k(U)$ with $|g - f|_K^k < \varepsilon$. These ε-neighbourhoods give $C_K^k(U)$ a topology and hence open, closed, dense, etc., have a meaning. Let

$$C_K^k(U, \mathbf{R}^p) = \underbrace{C_K^k(U) \times \ldots \times C_K^k(U)}_{p \text{ factors}},$$

with the maximum norm of the components for its norm.

The last assertion says:

2.15. <u>Let</u> $K \subset U \subset \mathbf{R}^n$, K <u>compact</u>, U <u>open</u>, $p \geq 2n$. <u>The set</u> ϑ <u>of differentiable maps</u> $f : U \to \mathbf{R}^p$, <u>such that</u> $Rk_x f = n$ <u>for</u> $x \in K$, <u>is open and dense in</u> $C_K^k(U, \mathbf{R}^p)$ <u>for each</u> k.

Proof. If $f : U \to \mathbf{R}^p$ is an immersion, $K \subset U$, then consider the map

$$U \xrightarrow{Df} \mathbf{R}^{n.p} \xrightarrow{\phi} \mathbf{R}$$

where $\phi(A)$ = the sum of the squares of the $(n \times n)$-minors of the matrix A.

The composite $\phi \circ Df$ does not vanish anywhere on K by hypothesis. If $f_1 : U \to \mathbf{R}^p$ is differentiable and Df_1 is sufficiently near Df on K, then $\phi \circ Df_1 \neq 0$ everywhere on K. Hence the set ϑ is open. It is dense because one can find a linear map A, which is arbitrarily small on K, such that $f + A$ is an immersion. √

The topologist states this result by saying that a map $\mathbf{R}^n \supset U \to \mathbf{R}^p$, $p \geq 2n$, is <u>almost always</u> an immersion.

3 · Construction of differentiable maps

The great freedom that one has in the construction of differentiable maps with pre-assigned properties, is based on the following fact:

3.1. **Lemma.** Let $\lambda : \mathbf{R} \to \mathbf{R}$ be defined by

$$\lambda(t) = 0 \text{ for } t \leq 0$$
$$\lambda(t) = e^{-1/t} \text{ for } t > 0$$

then $0 \leq \lambda(t) \leq 1$, and λ is (arbitrarily often) differentiable everywhere.

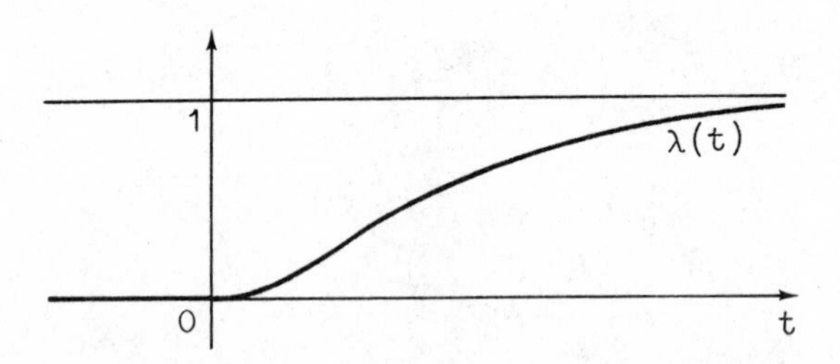

Proof. The n-th derivative of λ for $t > 0$ has the form $q(1/t) \cdot e^{-1/t}$, where q is a polynomial of degree $2n$. The derivatives converge to zero with t. Hence λ is differentiable at the origin and the Taylor series of λ at the origin is zero. √

Next let $\varepsilon > 0$ and $\phi_\varepsilon : \mathbf{R} \to \mathbf{R}$ be defined by

$$\phi_\varepsilon(t) = \frac{\lambda(t)}{\lambda(t) + \lambda(\varepsilon - t)}$$

ϕ_ε is differentiable, $0 \leq \phi_\varepsilon \leq 1$ and $\phi_\varepsilon(t) = 0$ iff $t \leq 0$, $\phi_\varepsilon(t) = 1$ for $t \geq \varepsilon$

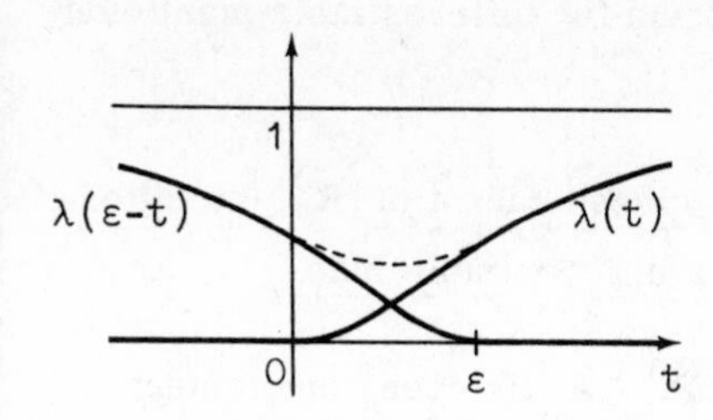

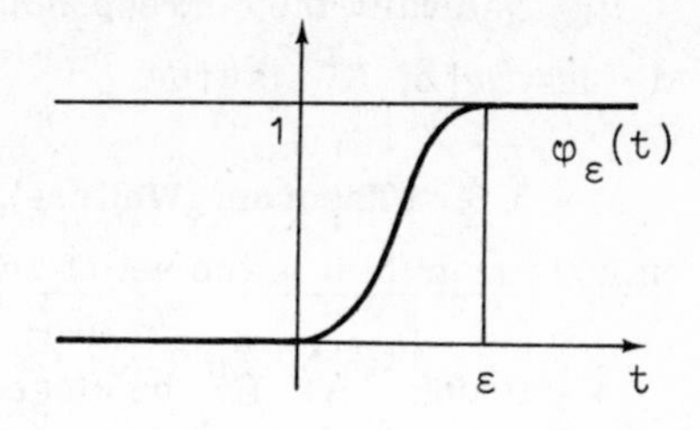

If $K(x, r) = \{y \in \mathbf{R}^n \,|\, |y - x| \le r\}$ is the ball around x with radius r, and $\varepsilon > 0$, then the function

$$\psi : \mathbf{R}^n \to \mathbf{R}$$
$$y \mapsto 1 - \phi_\varepsilon(|y - x| - r), \quad \text{for } x, r, \varepsilon \text{ fixed}$$

has the following properties:

$0 \le \psi(y) \le 1$, $\psi(y) = 1$ for $y \in K(x, r)$ and ψ is differentiable since at the points where $|y - x|$ is not differentiable, ψ is locally constant. Also $\psi(y) = 0$ iff $y \notin \mathring{K}(x, r+\varepsilon)$.

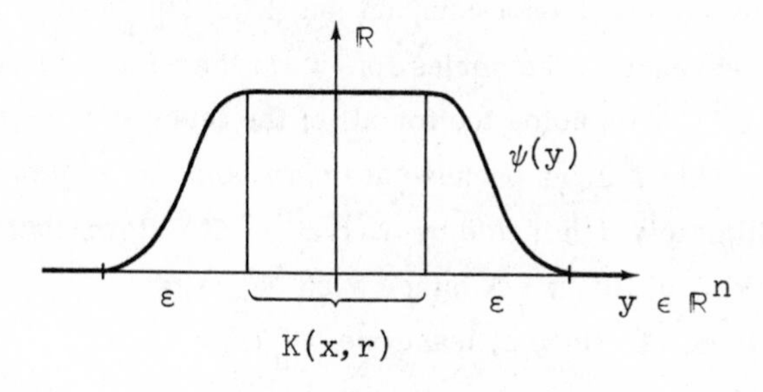

First application:

3.2. Let $C^\infty(\mathbf{R}^n)$ be the set of differentiable functions $\mathbf{R}^n \to \mathbf{R}$, $\mathcal{E}_x(n)$ the set of germs: $(\mathbf{R}^n, x) \to \mathbf{R}$, and $p_x : C^\infty(\mathbf{R}^n) \to \mathcal{E}_x(n)$ the map which assigns to a map f its germ at x, then p_x is surjective.

Proof. Let $\tilde{\phi} : (\mathbf{R}^n, x) \to \mathbf{R}$ be represented by $\phi : U \to \mathbf{R}$, choose $r, \varepsilon > 0$ such that $K(x, r+\varepsilon) \subset U$, and choose ψ as above. Then $\tilde{\phi} = p_x(\phi.\psi)$ and $(\phi.\psi)(y) = 0$ outside $K(x, r+\varepsilon)$ so that $\phi.\psi$ can be extended to all of $\mathbf{R}^n$ using the zero function. ✓

Naturally the corresponding theorem for differentiable manifolds M^n instead of $\mathbf{R}^n$ is true.

3.3. **Theorem (Whitney).** Any closed subset of $\mathbf{R}^n$ (or differentiable manifold) is the set of zeros of a differentiable map.

Proof. $A \subset \mathbf{R}^n$ be closed, $U = \mathbf{R}^n - A$ its open complement. Without loss $U \neq \emptyset$, so that $U = \bigcup_{m \in \mathbf{N}} \mathring{K}(x_m, r_m)$. Let $\psi_m : \mathbf{R}^n \to \mathbf{R}$ be differentiable with $\psi_m \geq 0$ and

$$\psi_m(y) \neq 0 \iff y \in \mathring{K}(x_m, r_m).$$

Let $\psi : \mathbf{R}^n \to \mathbf{R}$ be defined by

$$\psi(y) = \sum_{m=1}^{\infty} \psi_m(y) \cdot \varepsilon_m$$

where $\{\varepsilon_m | m \in \mathbf{N}\}$ is a real sequence, each $\varepsilon_m > 0$, chosen so that each derivative of $\varepsilon_m \psi_m$ of order $\leq m$ (thus finitely many such derivatives for given m) has absolute value $\leq 1/2^m$. Since ψ_m and its derivatives are only non-zero on a compact set $K(x_m, r_m)$, it is possible to find such a sequence. The series for ψ is therefore uniformly convergent on all $\mathbf{R}^n$. This holds too for all of the term by term derivatives of this series. This follows because at every point of $\mathbf{R}^n$ any of these series is ultimately dominated by $\Sigma(1/2^m)$. It follows that ψ is differentiable, $\psi(x) = 0$ for $x \in A$ since each $\psi_m(x) = 0$ for $x \in A$, and $\psi(x) \neq 0$ outside A since at least one $\psi_m(x) \neq 0$. ✓

One should see this result in relation to Sard's theorem: a set defined by continuous equations is closed and any closed set can be described by differentiable equations. Now Sard's theorem said it is not likely that the solution set of $f(x) = b$ will be pathological, for differentiable f.

3.4. **Exercises.** 1. Let A_0, A_1 be disjoint closed sets in $\mathbf{R}^n$, prove there exists a differentiable function $\phi : \mathbf{R}^n \to \mathbf{R}$, $0 \leq \phi(x) \leq 1$, with $A_0 = \phi^{-1}\{0\}$; $A_1 = \phi^{-1}\{1\}$.

2. Let $A \subset \mathbf{R}^n$ be closed, prove there is a differentiable map $f : \mathbf{R}^n \to \mathbf{R}^n$, such that A is the set of critical points of f.

3. There is a differentiable map $f : \mathbf{R} \to \mathbf{R}^n$ such that $f(\mathbf{R})$ is dense in $\mathbf{R}^n$.

3.5. **Definition.** Let $\mathbf{R}^n \supset A \xrightarrow{f} \mathbf{R}^k$, then

$$\mathrm{Supp}(f) = \overline{\{x \in A \mid f(x) \neq 0\}} \quad \text{(Closure in } A),$$

the support (or carrier) of f, i.e. $x \notin \mathrm{Supp}(f)$ iff the germ $\tilde{f}: (A, x) \to \mathbf{R}^k$ vanishes.

3.6. **Theorem** (partition of unity). Let M be a differentiable manifold and $\{U_\lambda \mid \lambda \in \Lambda\}$ an open cover of M, then there is a sequence of differentiable functions $\{\phi_n \mid n \in \mathbf{N}\}$ such that

$$\phi_n : M \to \mathbf{R}, \quad 0 \leq \phi_n(x) \leq 1$$

$$\mathrm{Supp}(\phi_n) \subset U_{\lambda(n)} \text{ for some } \lambda(n) \in \Lambda$$

$$\{\mathrm{Supp}(\phi_n) \mid n \in \mathbf{N}\} \text{ is locally finite}$$

and $\sum_{n \in \mathbf{N}} \phi_n(x) = 1$, for all $x \in M$.

3.7. Locally finite means that each point $x \in M$ has a neighbourhood V, such that $V \cap \mathrm{Supp}(\phi_n) = \emptyset$ for all but finitely many $n \in \mathbf{N}$.

The set $\{\phi_n \mid n \in \mathbf{N}\}$ is called a partition of unity associated with the cover $\{U_\lambda \mid \lambda \in \Lambda\}$.

The sum $\Sigma_{n \in \mathbf{N}} \phi_n$ is well defined, because near any point only a finite number of the ϕ_n are non-zero. Typical application:- Choose the U_λ to be balls in the charts of a coordinate system. The closures $\overline{U}_\lambda$ will be compact and so $\mathrm{Supp}(\phi_n)$ will be compact. If $f : M \to \mathbf{R}^k$ is any map, then

$$f = \sum_{n \in \mathbf{N}} f_n \text{ is locally finite, where}$$

$$f_n = \phi_n \cdot f, \text{ hence}$$

$$f_n = 0 \text{ outside the compact set } \mathrm{Supp}(\phi_n).$$

Proof of the theorem. M is locally compact with a countable basis, hence (Schubert, Topologie, p. 90) one may find a cover $\{V_n \mid n \in \mathbf{N}\}$,

$\overline{V}_n \subset U_{\lambda(n)}$, so that $\{\overline{V}_n | n \in \mathbf{N}\}$ is locally finite and without loss of generality each $\overline{V}_n$ is contained in a coordinate neighbourhood. Using Whitney's theorem, one obtains a positive differentiable function $\tau_n : M \to \mathbf{R}$ with $\tau_n(x) = 0$ iff $x \notin V_n$. Clearly $\mathrm{Supp}(\tau_n) = \overline{V}_n$. Define

$$\tau(x) = \sum_{n \in \mathbf{N}} \tau_n(x).$$

$\tau(x) > 0$ for all x and τ is differentiable since the sum is locally finite. Hence we may define $\phi_n = \tau_n/\tau$. √

Later we shall often say:

'without loss, Supp(f) is compact and contained in a ball with radius $< \varepsilon$, ...'.

This is always based on the fact that any f is a locally finite sum of such functions.

4 · Germs and jets

Literature: R. Narasimhan: Analysis on real and complex manifolds, Masson and Cie, Paris, and North-Holland, Amsterdam, (1968).

S. Lang: Algebra, Addison-Wesley (1969).

J. Mather: Stability of C^∞-mappings III, I. H. E. S. Publ. Math., 35 (1968), 127-56.

N. Bourbaki: Algèbre, Ch. IV.

Let

$$\mathcal{E}(n) = \text{the ring of differentiable germs: } (\mathbf{R}^n, 0) \to \mathbf{R}.$$

If $C^\infty(n)$ = the ring of differentiable maps: $\mathbf{R}^n \to \mathbf{R}$, then there is a surjective map: $C^\infty(n) \to \mathcal{E}(n)$, $f \mapsto \tilde{f}$. In $C^\infty(n)$ there is the ideal $\mathfrak{a}$:

$$\mathfrak{a} = \{f \in C^\infty(n) \mid f \text{ vanishes on a neighbourhood of } 0\}.$$

As rings $\mathcal{E}(n) = C^\infty(n)/\mathfrak{a}$ and this can be used to define the ring structure on $\mathcal{E}(n)$.

Let

$$\mathfrak{m}(n) = \{f \in \mathcal{E}(n) \mid f(0) = 0\}.$$

Then $\mathfrak{m}(n) \subset \mathcal{E}(n)$ is a maximal ideal and $\mathcal{E}(n)/\mathfrak{m}(n) = \mathbf{R}$, given by $\tilde{f} \mapsto \tilde{f}(0)$. In fact $\mathfrak{m}(n)$ is the unique maximal ideal in $\mathcal{E}(n)$. To see this, suppose $\tilde{f} \notin \mathfrak{m}(n)$, then $\tilde{f}(0) \neq 0$ and so on a neighbourhood of the origin, $f(x) \neq 0$ for any representative of $\tilde{f}$. It follows that $1/f$ is defined which means that $\tilde{f}$ is a unit and not contained in any proper ideal. Let $C^\infty(M) = \{f : M \to \mathbf{R} \mid f \text{ is differentiable}\}$.

4.1. Exercises. 1. Let M be a differentiable manifold. Prove that $\mathfrak{a}_x = \{f \in C^\infty(M) \mid f(x) = 0\} \subset C^\infty(M)$ is a maximal ideal.

2. If M is a compact differentiable manifold and $\mathfrak{a}$ is a maximal ideal in $C^\infty(M)$, show that there is an $x \in M$, such that $\mathfrak{a} = \{f \in C^\infty(M) \mid f(x) = 0\}$.

3. If M is not compact, show there is a maximal ideal $\mathfrak{a} \subset C^\infty(M)$, such that for each $x \in M$ there is an $f \in \mathfrak{a}$ with $f(x) \neq 0$.

4. Let $\alpha : \mathcal{E}(n) \to \mathcal{E}(k)$ be a homomorphism of rings. Show that $\alpha = 0$ or $\alpha(1) = 1$. Show also $\alpha(\mathfrak{m}(n)) \subset \mathfrak{m}(k)$.

Let $x_1, \ldots, x_n$ be coordinate functions, then $\mathfrak{m}(n)$ is generated by the germs $\tilde{x}_1, \ldots, \tilde{x}_n$. More generally:

4.2. Theorem. Let $\mathfrak{m}(k) \subset \mathcal{E}(n + k)$ be the ideal of germs $\tilde{f} : \mathbf{R}^n \times \mathbf{R}^k \to \mathbf{R}$ (at the origin), for which $\tilde{f} \mid \mathbf{R}^n \times \{0\} = 0$, and let $(x_1, \ldots, x_n, y_1, \ldots, y_k)$ be coordinates on $\mathbf{R}^n \times \mathbf{R}^k$, then $\mathfrak{m}(k)$ is generated by the germs $\tilde{y}_i$ (i.e. the germs of $(x, y) \mapsto y_i$). Hence

$$\tilde{f} \in \mathfrak{m}(k) \Longleftrightarrow \tilde{f} = \sum_{i=1}^{k} \tilde{y}_i \tilde{f}_i \quad \text{with} \quad \tilde{f}_i \in \mathcal{E}(n + k).$$

Proof. Choose a representative $f : \mathbf{R}^n \times \mathbf{R}^k \to \mathbf{R}^k$ with $f \mid \mathbf{R}^n \times \{0\} = 0$, then

$$\begin{aligned} f(x, y) &= \int_0^1 \frac{d}{dt} f(x, ty)dt \\ &= \int_0^1 \Big(\sum_{i=1}^{k} \frac{\partial f}{\partial y_i}(x, ty).\, y_i\Big)dt \\ &= \sum_{i=1}^{k} y_i \,.\, \underbrace{\int_0^1 \frac{\partial f}{\partial y_i}(x, ty)dt}_{f_i(x,\, y)} \\ &= \sum_{i=1}^{k} y_i .\, f_i(x, y), \quad f_i \text{ differentiable.} \quad \checkmark \end{aligned}$$

This theorem has a number of consequences. $\mathcal{E}(n)$ is an $\mathbf{R}$-algebra, a derivation of $\mathcal{E}(n)$ is a linear map

$$X : \mathscr{E}(n) \to \mathbf{R}, \text{ such that}$$
$$X(f.g) = X(f).g(0) + f(0).X(g).$$

In particular $X(1) = X(1.1) = X(1) + X(1)$, hence $X(1) = 0$ and $X(c) = 0$ for constant functions.

The collection of derivations forms a vector space, furthermore:

4.3. Theorem. The maps $\partial/\partial x_i|_0 : \mathscr{E}(n) \to \mathbf{R}$, $\tilde{f} \mapsto \partial f/\partial x_i(0)$ form a basis of the vector space of derivations of $\mathscr{E}(n)$.

Proof. Linearly independent:- Assume $\lambda_i \in \mathbf{R}$ and

$$\sum_{i=1}^{n} \lambda_i \frac{\partial}{\partial x_i}\Big|_0 = 0,$$

then

$$\sum_{i=1}^{n} \lambda_i \frac{\partial \tilde{x}_j}{\partial x_i}\Big|_0 = \lambda_i = 0.$$

Spanning set:- Let X be a derivation and $X(x_i) = \lambda_i$. Then $Y = X - \sum_{i=1}^{n} \lambda_i \frac{\partial}{\partial x_i}\Big|_0$ is a derivation and $Y(x_i) = 0$. If $\tilde{f} \in \mathscr{E}(n)$ is written as $\tilde{f} = f(0) + \Sigma \tilde{x}_i \tilde{f}_i$, then

$$\begin{aligned} Y(\tilde{f}) &= Y(f(0)) + \sum_i Y(\tilde{x}_i . \tilde{f}_i) \\ &= 0 + \Sigma\, Y(\tilde{x}_i).\tilde{f}_i(0) + \Sigma\, \tilde{x}_i(0).\, Y(\tilde{f}_i) \\ &= 0. \end{aligned}$$

Hence

$$X = \sum_{i=1}^{n} \lambda_i \frac{\partial}{\partial x_i}\Big|_0 . \quad \checkmark$$

4.4. Notation. Let $\alpha = (\alpha_1, \ldots, \alpha_n)$ (resp. $(\alpha, \beta) = (\alpha_1, \ldots, \alpha_n, \beta_1, \ldots, \beta_k)$) $\alpha_i, \beta_k \in \mathbf{N} \cup \{0\}$. Then

$$\alpha! = \alpha_1! \ldots \alpha_n!$$
$$0! = 1$$
$$|\alpha| = \alpha_1 + \ldots + \alpha_n$$

$$D^{\alpha} f = \frac{\partial^{|\alpha|}}{\partial x_1^{\alpha_1} \dots \partial x_n^{\alpha_n}} f,$$

$$D^{\alpha,\beta} f = \frac{\partial^{|\alpha|+|\beta|}}{\partial x_1^{\alpha_1} \dots \partial x_n^{\alpha_n} \partial y_1^{\beta_1} \dots \partial y_k^{\beta_k}} f.$$

$|\alpha|$ = the order of D^{α}

$$(x, y)^{\alpha,\beta} = x_1^{\alpha_1} . x_2^{\alpha_2} \dots x_n^{\alpha_n} . y_1^{\beta_1} \dots y_k^{\beta_k}$$

4.5. Theorem. Let $\mathfrak{m}(k) \subset \mathcal{E}(n+k)$ as in theorem 4.2, then $\mathfrak{m}(k)^s = \{\tilde{f} \in \mathcal{E}(n+k) \mid D^{\alpha,\beta} f | \mathbf{R}^n \times \{0\} = 0$ for all α and all β with $|\beta| < s\}$, and this ideal is generated by the monomials

$$y_1^{\beta_1} \dots y_k^{\beta_k}, \quad |\beta| = s.$$

Proof. $\mathfrak{m}(k)^s = \{f \in \mathcal{E}(n+k) \mid f = \sum_{\lambda} f_{\lambda_1} \dots f_{\lambda_s}$ with $f_{\lambda_i} \in \mathfrak{m}(k)\}$, and the second assertion follows immediately. From the product rule

$$f \in \mathfrak{m}(k)^s \Rightarrow D^{\alpha,\beta} f | \mathbf{R}^n \times \{0\} = 0 \quad \text{for} \quad |\beta| < s.$$

If, on the other hand, $D^{0,\beta} f | \mathbf{R}^n \times \{0\} = 0$ for all $|\beta| < s$, then $f \in \mathfrak{m}(k)^{s-1}$ (by induction) and $f = \sum_{|\beta|=s-1} f_\beta y^\beta$. It suffices to show that $f_\beta \in \mathfrak{m}(k)$, i.e. $f_\beta | \mathbf{R}^n \times \{0\} = 0$ for all β, $|\beta| = s - 1$. If some $f_{\beta_0} | \mathbf{R}^n \times \{0\} \neq 0$, then $D^{0,\beta_0} f = \sum_\beta D^{0,\beta_0} f_\beta . y^\beta = \beta_0! . f_{\beta_0} \neq 0$ on $\mathbf{R}^n \times \{0\}$. ✓

4.6. Consider the following diagram (note $\mathfrak{m}^k \subset \mathfrak{m}^l$ for $k \geq l$)

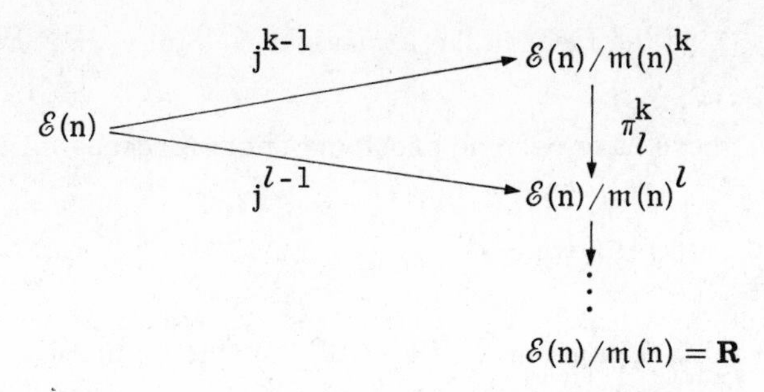

The image $j^{k-1}(f)$ of $\tilde{f} \in \mathcal{E}(n)$ is called the (k-1)-jet of $\tilde{f}$, sometimes denoted $\hat{f}$. The quotient $\mathcal{E}(n)/\mathfrak{m}(n)^k$ is the **R**-algebra of (k-1)-jets.

Two germs define (have) the same k-jet at the origin of $\mathbf{R}^n$, if their derivatives up to the k-th order are identical. Clearly $j^k(f) = j^k(\sum_{|\alpha| \leq k} \frac{D^\alpha f(0)}{\alpha!} x^\alpha)$, (the k-th Taylor polynomial), and two polynomials of degree $\leq k$ have the same k-jet if and only if they are identical.

4.7. A germ $\tilde{f} \in \mathfrak{m}(n)^k$ is said to vanish to order k (it has zero (k-1)-jet). The symbol o(k) will sometimes be used to denote a germ which vanishes to order k.

We see that any k-jet is represented by a polynomial of degree $\leq k$. These polynomials are added in the usual way and they are multiplied in the usual way except that terms of order $>k$ are omitted from the product. In particular:

$$\mathcal{E}(n)/\mathfrak{m}(n)^{k+1} = \mathbf{R}[x_1, \ldots, x_n]/\langle x_1, \ldots, x_n\rangle^{k+1}$$

where $\mathbf{R}[x_1, \ldots, x_n]$ is the ring of polynomials in $x_1, \ldots, x_n$ and $\langle x_1, \ldots, x_n\rangle^{k+1}$ is the ideal generated by $\{x_1, \ldots, x_n\}$ raised to the power $(k + 1)$.

Also $j^k(f) =$ k-th Taylor polynomial of f at $0 \in \mathbf{R}^n$.

4.8. More generally:

$$\mathcal{E}(n+k)/\mathfrak{m}(k)^{s+1} = \mathcal{E}(n)[y_1, \ldots, y_k]/\langle y_1, \ldots, y_k\rangle^{s+1}$$

$$\tilde{f} \mapsto \sum_{|\beta| \leq s} (D^{0,\beta} f | \mathbf{R}^n \times \{0\}) \cdot \frac{y^\beta}{\beta!}$$

where $\langle y_1, \ldots, y_k \rangle$ is the ideal in the $\mathcal{E}(n)$-algebra $\mathcal{E}(n)[y_1, \ldots, y_k]$ generated by $\{y_1, \ldots, y_k\}$.

Up till now, there have been no problems of convergence. Define

$$\mathfrak{m}(k)^{\infty} = \bigcap_{s=1}^{\infty} \mathfrak{m}(k)^s \subset \mathcal{E}(n+k)$$

Using theorem 4.5, a germ $f : (\mathbf{R}^n \times \mathbf{R}^k, 0) \to \mathbf{R}$ is in $\mathfrak{m}(k)^{\infty}$ if and only if, for arbitrarily large $s \in \mathbf{N}$, there is a representation $f = \sum_{|\beta|=s} f_\beta y^\beta$. That is, if and only if all germs $D^{\alpha,\beta} f$ vanish on $\mathbf{R}^n \times \{0\}$.

4.9. Theorem (Borel). Let $\mathfrak{m}(k)^{\infty} \subset \mathcal{E}(n+k)$ be defined as above, then

$$\mathcal{E}(n+k)/\mathfrak{m}(k)^{\infty} = \mathcal{E}(n)[[y_1, \ldots, y_k]]$$

$$f \mapsto \sum_\beta (D^{0,\beta} f | \mathbf{R}^n \times \{0\}) \cdot \frac{y^\beta}{\beta!} .$$

The ring on the right is the ring of formal power series in the variables $y_1, \ldots, y_k$ over the ring $\mathcal{E}(n)$. For $n = 0$, this gives

$$\mathcal{E}(k)/\mathfrak{m}(k)^{\infty} = \mathbf{R}[[x_1, \ldots, x_k]],$$

that is, for any given power series (not necessarily convergent) there exists a function whose Taylor series at the origin is exactly this power series.

Proof. The map $\mathcal{E}(n+k)/\mathfrak{m}(k)^{\infty} \to \mathcal{E}(n)[[y_1, \ldots, y_k]]$ is obviously injective, for if $f \in \mathcal{E}(n+k)$ and $D^{0,\beta} f | \mathbf{R}^n \times \{0\} = 0$ for all β, then by theorem 4.5, $f \in \mathfrak{m}(k)^s$ for all s. It remains to prove:

4.10. For each $\beta = (\beta_1, \ldots, \beta_k)$ let the germ $\tilde{f}_\beta : (\mathbf{R}^n, 0) \to \mathbf{R}$ be given, then there exists a germ $\tilde{f} : (\mathbf{R}^n \times \mathbf{R}^k, 0) \to \mathbf{R}$, with $(D^{0,\beta} \tilde{f} | \mathbf{R}^n \times \{0\}) = \tilde{f}_\beta$.

Choose representatives $f_\beta : \mathbf{R}^n \to \mathbf{R}$ for the $\tilde{f}_\beta$ with compact supports in $K(0, 1)$. Take a function $\phi : \mathbf{R}^k \to \mathbf{R}$, $0 \le \phi \le 1$, $\phi(y) = 1$ for $|y| \le 1/2$, $\phi(y) = 0$ for $|y| \ge 1$. Let $y = (y_1, \ldots, y_k)$ and

$$(*) \qquad f(x, y) = \sum_\beta \frac{f_\beta(x)}{\beta!} \cdot y^\beta \cdot \phi(t_\beta \cdot y), \; 1 < t_\beta \in \mathbf{R}.$$

Assume that the sequence $t_\beta = t_{|\beta|}$ can be defined to make the series

$$(**) \qquad \sum_\beta D^\alpha\left(\frac{f_\beta(x)}{\beta!} \cdot y^\beta \cdot \phi(t_\beta \cdot y)\right)$$

uniformly convergent for each $\alpha = (\alpha_1, \ldots, \alpha_{n+k})$. Then f would be well defined and differentiable and (*) could be differentiated term by term. Since the germ of $\phi(t_\beta \cdot y)$ is equal to 1, this gives

$$D^{0,\beta} f | \mathbf{R}^n \times \{0\} = \tilde{f}_\beta$$

as required. Therefore it has to be proved that a sufficiently rapidly increasing sequence $\{t_{|\beta|}\}$ makes the series (**) uniformly convergent for every α.

For this, write the β-th term in (*) in the form $(t_{|\beta|} > 1)$

$$(1/t_{|\beta|})^{|\beta|} \cdot \frac{f_\beta(x)}{\beta!} \cdot (t_{|\beta|} \cdot y)^\beta \cdot \phi(t_{|\beta|} \cdot y)$$

$$= (1/t_{|\beta|})^{|\beta|} \cdot f_\beta(x) \cdot \psi_\beta(t_{|\beta|} \cdot y)$$

The functions ψ_β vanish outside $\{ |t_{|\beta|} \cdot y| \le 1 \}$. Next let

$$M_\beta = \max\{ |D^\alpha(f_\beta(x) \cdot \psi_\beta(y))|, \; |\alpha| < |\beta| \}.$$

Observe that $\mathrm{Supp}(f_\beta \cdot \psi_\beta) \subset \{(x, y) \mid |x|, |y| \le 1\}$, and that there are only finitely many α with $|\alpha| < |\beta|$. Hence M_β exists.

Since $t_{|\beta|} > 1$, it follows for $|\alpha| < |\beta|$ that
$|\beta\text{-th element in } (**)| \le (t_{|\beta|})^{|\alpha|} \cdot (1/t_{|\beta|})^{|\beta|} \cdot M_\beta < M_\beta/t_{|\beta|}$
(remember $\alpha = (\alpha_1, \ldots, \alpha_{n+k})$). Choose a sequence $\varepsilon_\beta > 0$, such that $\sum_\beta \varepsilon_\beta$ converges and choose $t_{|\beta|} > M_\beta/\varepsilon_\beta$. Finally, for $|\beta| > |\alpha|$ the β-th element of (**) is dominated by ε_β. ✓

As has been said, the formal power series form a ring $\mathbf{R}[[x_1, \ldots, x_n]]$. This will also be denoted $\hat{\mathcal{E}}(n)$, with elements $\hat{f}, \hat{g}, \ldots$.

Hence the map in Borel's theorem

$$\mathscr{E}(n) \xrightarrow{j=j^\infty} \mathscr{E}(n)/m(n)^\infty = \hat{\mathscr{E}}(n)$$

can also be indicated by $\tilde{f} \mapsto \hat{f}$. If

$$\hat{f} = \Sigma_\alpha f_\alpha x^\alpha \qquad f_\alpha, g_\alpha \in \mathbf{R},$$
$$\hat{g} = \Sigma_\alpha g_\alpha x^\alpha$$

then

$$\hat{f} + \hat{g} = \Sigma_\alpha (f_\alpha + g_\alpha) x^\alpha; \ \hat{f} . \hat{g} = \Sigma_\alpha (\sum_{\beta+\gamma=\alpha} f_\beta . g_\gamma) x^\alpha .$$

If $\tilde{f} \in \mathscr{E}(n)$, then $j(f) = j^\infty(f) = \hat{f}$ is the jet (or ∞-jet) of $\tilde{f}$ at the origin.

4.11. The map $\mathscr{E}(n) \to \hat{\mathscr{E}}(n)$ is a homomorphism of algebras. To see, for example, that

$$(f. g)\hat{} = \hat{f} . \hat{g}$$

up to order k (for arbitrary k), write

$$\hat{f} = p + m, \ \hat{g} = q + r, \ p, q \text{ polynomials}, \ m, r \in m(n)^{k+1},$$

then $\hat{f}. \hat{g} = p. q + m_1, \ m_1 \in m(n)^{k+1}$. Hence

$$(f. g)\hat{} = (p. q)\hat{} = p. q \bmod m(n)^{k+1}$$

since the Taylor expansion leaves polynomials unchanged.

4.12. $\hat{\mathscr{E}}(n)$ has a unique maximal ideal

$$\hat{m}(n) = \{\hat{f} \in \hat{\mathscr{E}}(n) \,|\, \hat{f}(0) = 0\}.$$

If $\hat{f} \notin \hat{m}(n)$, then $\hat{f} = f_0 . (1 - f_1)$, $f_1 \in \hat{m}(n)$, $0 \neq f_0 \in \mathbf{R}$, and this gives the power series

$$\frac{1}{\hat{f}} = \frac{1}{f_0} . (1 + f_1 + f_1^2 + \dots)$$

(with the monomials put into increasing order).

Hence $\hat{f} \notin \hat{m}(n) \Rightarrow \hat{f}$ is a unit ($\Rightarrow \tilde{f}$ is a unit).

4.13. The ideal $\hat{\mathfrak{m}}(n)$ is generated by $x_1, \ldots, x_n$ (each monomial is divisible by some x_i). Also

$$\hat{\mathfrak{m}}(n)^{\infty} = \bigcap_{s=1}^{\infty} \hat{\mathfrak{m}}(n)^s = \{0\},$$

since a power series $\hat{f} \neq 0$ has a degree equal to the smallest $k \in N$ such that

$$\hat{f} = \sum_{\beta} f_{\beta} x^{\beta} \text{ and } f_{\beta} \neq 0 \text{ for } |\beta| = k.$$

Clearly

$$\text{degree}(\hat{f}.\hat{g}) = \text{degree}(\hat{f}) + \text{degree}(\hat{g})$$

(degree $(0) = \infty$ by definition).

A power series $\hat{f} \in \hat{\mathfrak{m}}(n)^k$ has degree $\geq k$ so that $\hat{f} \in \hat{\mathfrak{m}}(n)^{\infty}$ implies degree $(\hat{f}) \geq k$ for all k, i.e. $\hat{f} = 0$.

The following result is less trivial:

4.14. $\hat{\mathcal{E}}(n)$ is Noetherian and factorial (unique factorisation domain) (see Bourbaki!).

Both of these properties are false for $\mathcal{E}(n)$. In fact the ideal $\mathfrak{m}(n)^{\infty} \subset \mathcal{E}(n)$ is not finitely generated. (This is left as an exercise; it is not completely trivial, since $\hat{\mathfrak{m}}(n)^{\infty} = 0$ and so is finitely generated, but it is easily solved using the next theorem.)

4.15. **Theorem (Nakayama-lemma).** Let $\mathcal{R}$ be a commutative ring with 1, which has a unique maximal ideal $\mathfrak{m}$. Suppose A is a finitely generated $\mathcal{R}$-module. Then $\mathfrak{m}.A = A \Rightarrow A = 0$.

Consequence. With the same assumptions, let B, C be $\mathcal{R}$-modules such that $A, B \subset C$ then

$$A \subset B + \mathfrak{m}.A \Rightarrow A \subset B.$$

Proof. Consequence: $A \subset B + \mathfrak{m} A$ implies

$$A/A \cap B \subset (B + \mathfrak{m} A)/B = \mathfrak{m}(A/A \cap B).$$

By Nakayama, $A/A \cap B = 0$ and so

$$A = A \cap B \text{ or } A \subset B. \quad \checkmark$$

Proof of theorem. If $z \in \mathfrak{m}$, then $1 + z$ is a unit (this is all that is used). For, otherwise $1 + z \in \mathfrak{m}$ ($1 + z$ must lie in some maximal ideal), which means $1 \in \mathfrak{m}$. If $a_1, \ldots, a_n$ are generators of A, then by hypothesis there are $z_{ij} \in \mathfrak{m}$ such that

$$a_i = \sum_{j=1}^{n} z_{ij} a_j.$$

As matrices this is $a = Za$, i.e. $(Z - 1)\, a = 0$, where 1 is the identity matrix.

Now, $\det(Z - 1)$ is the characteristic polynomial of Z at the point 1 and this is equal to (± 1 + sums of products of elements of Z)

$$= \pm 1 + \bar{z}, \quad \bar{z} \in \mathfrak{m}.$$

Hence $\det(Z - 1)$ is invertible, and so $Z - 1$ is invertible, hence $a = (a_1, \ldots, a_n) = 0$, or, $A = 0$. $\checkmark$

If $\mathcal{R} = \hat{\mathcal{E}}(n)$, then one may take A to be any ideal because the ring is Noetherian. However not every ideal in $\mathcal{E}(n)$ is finitely generated.

Returning to the rings that we have been studying, let

$$\begin{aligned} \mathcal{E}(n, p) &= \text{Ring of germs: } (\mathbf{R}^n, 0) \to \mathbf{R}^p \\ &= \mathcal{E}(n) \times \mathcal{E}(n) \times \ldots \times \mathcal{E}(n) \quad (p \text{ factors}), \end{aligned}$$

and correspondingly $\hat{\mathcal{E}}(n, p) = \hat{\mathcal{E}}(n) \times \ldots \times \hat{\mathcal{E}}(n)$.

One has the jet map $j : \mathcal{E}(n, p) \to \hat{\mathcal{E}}(n, p)$

$$j(\tilde{f}_1, \ldots, \tilde{f}_p) = (\hat{f}_1, \ldots, \hat{f}_p).$$

Given germs

$$(\mathbf{R}^n, 0) \xrightarrow{\tilde{f}} (\mathbf{R}^p, 0) \xrightarrow{\tilde{g}} \mathbf{R}^q,$$

then $\tilde{g} \circ \tilde{f} : (\mathbf{R}^n, 0) \to \mathbf{R}^q$ is a germ and correspondingly for the formal power series

$$\hat{f} = (\hat{f}_1, \ldots, \hat{f}_p),\ \hat{f}_i \in \hat{m}(n), \quad \text{and}$$

$$\hat{g} = (\hat{g}_1, \ldots, \hat{g}_q),\ \hat{g}_j \in \hat{\mathscr{E}}(p)$$

there is the composite $\hat{g} \circ \hat{f} \in \hat{\mathscr{E}}(n, q)$:

$$(\hat{g} \circ \hat{f})_i = \hat{g}_i(\hat{f}_1(x), \ldots, \hat{f}_p(x)).$$

The equality $(g \circ f)\hat{} = \hat{g} \circ \hat{f}$ is the <u>generalised chain rule.</u> (The proof is like the one for multiplication of jets.)

An element $\hat{f} \in \hat{\mathscr{E}}(n, p)$ may be differentiated term by term with respect to all the variables. The <u>Jacobian</u> $D\hat{f}(0)$ is given by the <u>linear part</u> of $\hat{f}$. There is an inverse function theorem.

4.16. Theorem. $\hat{f} \in \hat{\mathscr{E}}(n, p)$, $\hat{f}(0) = 0$ *is invertible with respect to '$\circ$' if and only if* $D\hat{f}(0)$ *is invertible (then* $n = p$). *The identity element for* $\hat{\mathscr{E}}(n, n)$ *under '$\circ$' is the n-tuple of power series* $(x_1, \ldots, x_n)$.

Proof. If $\hat{f} \circ \hat{g}(x) = x$, then $D\hat{f}(0) \,.\, D\hat{g}(0) = 1$, hence

$$\hat{f} \text{ invertible} \Rightarrow D\hat{f}(0) \text{ invertible}$$

(the Jacobian matrix is functorial!).

If $D\hat{f}(0)$ is invertible, choose $\tilde{f} \in \mathscr{E}(n, p)$ with $j(\tilde{f}) = \hat{f}$, so that $\tilde{f}^{-1} \circ \tilde{f}(x) = x$, $\tilde{f} \circ \tilde{f}^{-1}(y) = y$, hence

$$\hat{f}^{-1} \circ \hat{f}(x) = x, \quad \hat{f} \circ \hat{f}^{-1}(y) = y. \quad \checkmark$$

This proof should not be taken too seriously: the theorem is valid over arbitrary fields, and is much simpler than the inverse function theorem (see Bourbaki).

A differentiable germ $\tilde{f} : (\mathbf{R}^n, 0) \to (\mathbf{R}^p, 0)$ defines a homomorphism of algebras:

$$f^* : \mathscr{E}(p) \to \mathscr{E}(n)$$

$$\tilde{\phi} \mapsto \tilde{\phi} \circ \tilde{f}$$

and a jet $\hat{f} \in \hat{\mathscr{E}}(n, p)$, $\hat{f}(0) = 0$, defines a corresponding homomorphism

$$\hat{f}^* : \hat{\mathcal{E}}(p) \to \hat{\mathcal{E}}(n)$$

$$\hat{\phi}(y_1, \ldots, y_p) \mapsto \hat{\phi}(\hat{f}_1(x_1, \ldots, x_n), \ldots, \hat{f}_p(x_1, \ldots, x_n)).$$

The ring-homomorphism f^* makes $\mathcal{E}(n)$ into a module over $\mathcal{E}(p)$ as follows: if $\tilde{\phi} \in \mathcal{E}(p)$, $\tilde{\psi} \in \mathcal{E}(n)$, then let

$$\tilde{\phi} \,.\, \tilde{\psi} = f^*(\tilde{\phi}) \,.\, \tilde{\psi} = (\tilde{\phi} \circ \tilde{f}) \,.\, \tilde{\psi} \in \mathcal{E}(n),$$

similarly for $\hat{f}$.

The next two chapters will be devoted to the study of this module structure, in particular the connection between f^* and $\hat{f}^*$ and the question: for which f is the module $\mathcal{E}(n)$ finitely generated over $\mathcal{E}(p)$?

A simple preliminary is the following.

4.17. Remark. *Let $\tilde{f} : (\mathbf{R}^n, 0) \to (\mathbf{R}^p, 0)$ be a differentiable germ, then the following are equivalent*

(I) $\tilde{f}$ *is invertible*

(II) f^* *is an isomorphism*

(III) $\hat{f}^*$ *is an isomorphism.*

Proof. Clearly $(g \circ f)^* = f^* \circ g^*$ and $\mathrm{id}^* = \mathrm{id}$, so that $*$ is a functor, which transforms isomorphisms into isomorphisms. Hence (I) $\Rightarrow$ (II), (III).

Conversely the homomorphism of algebras $f^* : \mathcal{E}(p) \to \mathcal{E}(n)$ defines a homomorphism $d(f)$ from the vector space of derivations on $\mathcal{E}(n)$ to the vector space of derivations on $\mathcal{E}(p)$, given by

$$d(f)(X) = X \circ f^* : \mathcal{E}(p) \to \mathbf{R}.$$

Or, with respect to the canonical basis in this vector space

$$\begin{aligned} d(f)\left(\frac{\partial}{\partial x_i}\Big|_0\right)(\tilde{\phi}) &= \frac{\partial}{\partial x_i}\Big|_0 (f^*\tilde{\phi}) \\ &= \frac{\partial}{\partial x_i}\Big|_0 (\tilde{\phi} \circ \tilde{f}) \\ &= \sum_j \frac{\partial \phi}{\partial y_j} \cdot \frac{\partial f_j}{\partial x_i}(0) \end{aligned}$$

that is, $d(f) : \frac{\partial}{\partial x_i}\Big|_0 \mapsto \sum_j \frac{\partial f_j}{\partial x_i}(0) \cdot \frac{\partial}{\partial y_j}\Big|_0 .$

Hence the matrix of $d(f)$ with respect to the canonical bases for the vector spaces is the Jacobian matrix $(\frac{\partial f_j}{\partial x_i}(0))$. Now if f^* is an isomorphism, then $d(f)$ is an isomorphism, hence $D\tilde{f}$ is invertible and therefore $\tilde{f}$ is invertible.

Similarly one proves (III) $\Rightarrow$ (I). ✓

4.18. **Exercises.** 1. Let $f : \mathbf{R}^n \to \mathbf{R}$ be differentiable, and the germ $\tilde{f} : (\mathbf{R}^n, 0) \to \mathbf{R}$ of f an element of $\mathfrak{m}(n)^\infty$. Prove that the function $g : \mathbf{R}^n \to \mathbf{R}$, $g(x) = f(x)/|x|$, $g(0) = 0$, is differentiable.

2. Let $f : \mathbf{R} \to \mathbf{R}$ be differentiable and $f(x) = f(-x)$ for all $x \in \mathbf{R}$. Prove that there is precisely one differentiable function $g : \mathbf{R}_+ \to \mathbf{R}$ such that $f(x) = g(x^2)$.

3. Let $f : (\mathbf{R}, 0) \to (\mathbf{R}, 0)$ be a differentiable germ. Show that $f \in \mathfrak{m}(1)^k$, $f \notin \mathfrak{m}(1)^{k+1}$ if and only if there is an invertible germ $h : (\mathbf{R}, 0) \to (\mathbf{R}, 0)$ such that

$$f \circ h(x) = \pm x^k.$$

(This gives the complete classification of germs with non-vanishing jets under right transformations!)

4. Prove that if $\tilde{f} : (\mathbf{R}^n, 0) \to (\mathbf{R}^p, 0)$ is a germ, such that $f^* : \mathcal{E}(p) \to \mathcal{E}(n)$ is surjective, then $\tilde{f}$ is an immersion.

5 · The division theorem

Literature: L. Hörmander: Complex analysis in several variables, van Nostrand (1967).

J. Mather: Stability of C^∞-mappings I, Ann. of Math., 87 (1968), 89-104.

L. Nirenberg: A proof of the Malgrange-Mather preparation theorem, Proc. of Liverpool Singularities, Symposium I, Springer Lecture Notes 192 (1971), 97-105.

B. Malgrange: The preparation theorem for differentiable functions, Differential Analysis, Bombay Colloquium 1964, Oxford Univ. Press.

In order not to have to break off later for explanations, we begin with a few preliminaries which are not related to one another.

First Algebra:

Let $\{\alpha_i | i = 1, \ldots, n\}$ be indeterminates and $\{\sigma_i | i = 1, \ldots, n\}$ be defined by

$$\prod_{i=1}^{n} (x - \alpha_i) = \sum_{i=0}^{n} \sigma_i x^{n-i}, \text{ where } \sigma_0 = 1.$$

$(-1)^i \sigma_i(\alpha_1, \ldots, \alpha_n)$ is called the i-th elementary symmetric function.

Consider the map

$$\sigma : \mathbf{C}^n \to \mathbf{C}^n$$

$$(\alpha_1, \ldots, \alpha_n) \mapsto (\sigma_1, \ldots, \sigma_n).$$

5.1. **Lemma.** For each $\varepsilon > 0$, there exists a $\delta = \delta(\varepsilon) > 0$, such that $|\sigma_i| < \delta$ for $i = 1, \ldots, n \Rightarrow |x| < \varepsilon$ for x satisfying $\sum_{i=0}^{n} \sigma_i x^{n-i} = 0$. That is, if the coefficients of a polynomial converge to zero, then all the roots also converge to zero.

Proof. Let $p(x) = \sum_{i=0}^{n} \sigma_i x^{n-i} = x^n(1 + \frac{\sigma_1}{x} + \frac{\sigma_2}{x^2} + \ldots + \frac{\sigma_n}{x^n})$ (for $x \neq 0$). Now if $|x| \geq \varepsilon$ and the σ_i are sufficiently small (smaller than $\delta(\varepsilon)$) then the sum $\frac{\sigma_1}{x} + \ldots + \frac{\sigma_n}{x^n}$ is small and $(1 + \ldots) \neq 0$. Hence $p(x) \neq 0$. Thus, if $p(x) = 0$ and the σ_i are sufficiently small, then $|x| < \varepsilon$. ✓

In other words: if one assigns a root to each polynomial, then the map, taking the coefficients of the polynomial onto the root, is continuous at the origin, no matter which way the assignment of the root is made.

5.2. **Diversion.** The relation between the coefficients and the roots of a polynomial can be described as follows: For any topological space X, form the n-th <u>symmetric product</u>

$$SP^n(X) = \underbrace{X \times X \times \ldots \times X}_{n \text{ factors}}/\sim$$

where $(x_1, \ldots, x_n) \sim (x_{\pi(1)}, \ldots, x_{\pi(n)})$ for any permutation π of the numbers $(1, \ldots, n)$. Denote an equivalence class by $\prod_{i=1}^{n} x_i$.

<u>Fundamental theorem of Algebra</u> (for Topologists). <u>Let SP^n be the symmetric product and $S^2 = CP^1 = \{[a_i, b_i]\}$ in homogeneous complex coordinates. The map</u>

$$SP^n(S^2) \to CP^n$$

$$\prod_{i=1}^{n} [a_i, b_i] \mapsto [c_0, \ldots, c_n],$$

<u>where</u>

$$\prod_{i=1}^{n} (xa_i - yb_i) = \sum_{i=0}^{n} c_i y^i . x^{n-i}$$

<u>(well-defined independent of representatives) is a homeomorphism.</u>

Proof. Continuity is clear, the c_i are polynomials in the a_j, b_l. The map can also be written

$$\prod_{i=1}^{n} (xa_i - b_i) = \sum_{i=0}^{n} c_i x^{n-i}.$$

Injectivity means that the coefficients of a polynomial determine the roots. Surjectivity means that each polynomial over **C** decomposes into linear factors. Since $S^2 \times \ldots \times S^2$ is compact and $\mathbf{CP}^n$ is Hausdorff, the map from the quotient $SP^n(S^2)$ to $\mathbf{CP}^n$ is a homeomorphism: it is continuous and bijective. ✓

End of diversion.

5.3. **Exercise.** Prove that the set of $(\sigma_1, \ldots, \sigma_n) \in \mathbf{C}^n$, where the polynomial $\sum_{i=0}^{n} \sigma_i x^{n-i}$, $\sigma_0 = 1$ has fewer than n different complex roots, is closed and thin (the discriminant set).

5.4. If $\sigma \in \mathbf{C}^n$ is not in the discriminant set and α is a root of $\sum_{i=0}^{n} \sigma_i x^{n-i} = p_\sigma(x)$, then $p_\sigma(x) = (x - \alpha)g(x)$ and

$$\frac{\partial p_\sigma}{\partial x}(\alpha) = g(\alpha) \neq 0.$$

Hence the equation $\sum_{i=0}^{n} s_i x^{n-i} = 0$ can, in a neighbourhood of $s = \sigma$, $x = \alpha$, be solved for x with an analytic function. For,

$$\tau : (s_1, \ldots, s_n, x) \mapsto (s_1, \ldots, s_n, p_s(x))$$

is locally a coordinate transformation around $s = \sigma$, $x = \alpha$ and gives the function

$$(s_1, \ldots, s_n) \mapsto (\tau^{-1}(s_1, \ldots, s_n, 0))_{n+1}$$

which expresses a root of p_s as a function of the coefficients s (we have used the inverse function theorem for complex spaces).

The set $\{(\sigma, z) \in \mathbf{C}^n \times \mathbf{C} \,|\, p_\sigma(z) = 0\} \subset \mathbf{C}^n \times \mathbf{C}$ is always a submanifold of complex codimension 1, hence real codimension 2 (there is the corresponding result in the real case). This is because the set is also the graph of the function: $\mathbf{C}^{n-1} \times \mathbf{C} \to \mathbf{C}$ given by

$$\sigma_n(\sigma_1, \ldots, \sigma_{n-1}, z) = -\sum_{i=0}^{n-1} \sigma_i z^{n-i}.$$

The following example illustrates the real case in dimension 2 (alternatively the real part of the complex case):

5.6. **Example.** $n = 2$. $p_\sigma(z) = z^2 + 2\sigma_1 z + \sigma_2$ (for the sake of convenience). The discriminant set and the set $\{p_\sigma(z) = 0\}$ contain the origin of the coordinates in $\mathbf{C}^n \times \mathbf{C}$.

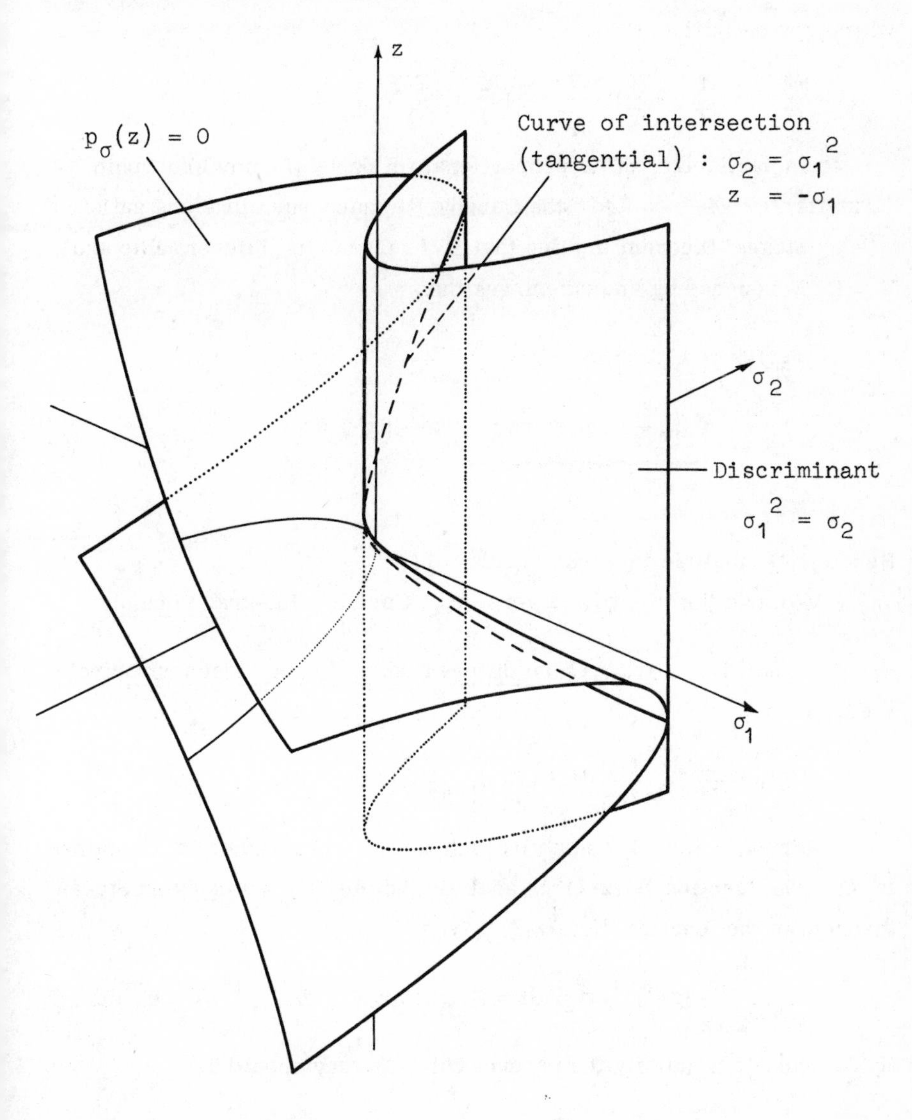

5.7. We need a little complex analysis. As real vector spaces: $\mathbf{C} \cong \mathbf{R}^2$. Let $f : \mathbf{C} \to \mathbf{C}$ be differentiable (as a real map). Take the usual coordinates $z = x + iy$; $\bar{z} = x - iy$:

$$x = \tfrac{1}{2}(z + \bar{z}); \quad y = -\tfrac{i}{2}(z - \bar{z}) .$$

The differential of f can be written

$$df = \frac{\partial f}{\partial x} dx + \frac{\partial f}{\partial y} dy = \frac{\partial f}{\partial z} dz + \frac{\partial f}{\partial \bar{z}} d\bar{z}$$

where, by definition,

$$\frac{\partial f}{\partial z} = \tfrac{1}{2}\left(\frac{\partial f}{\partial x} - i\frac{\partial f}{\partial y}\right); \quad \frac{\partial f}{\partial \bar{z}} = \tfrac{1}{2}\left(\frac{\partial f}{\partial x} + i\frac{\partial f}{\partial y}\right).$$

In particular, df is proportional to $dz \iff f$ is holomorphic (analytic) $\iff \partial f/\partial \bar{z} = 0 \iff$ the Cauchy-Riemann equations are satisfied.

Stokes' theorem implies that if $f : \mathbf{C} \to \mathbf{C}$ is differentiable and $D \subset \mathbf{C}$ is bounded by smooth curves then

$$\begin{aligned} \int_{\partial D} f dz &= \int_D df \wedge dz \\ &= \underbrace{\int_D \frac{\partial f}{\partial z} \cdot dz \wedge dz}_{0} + \int_D \frac{\partial f}{\partial \bar{z}} d\bar{z} \wedge dz. \end{aligned}$$

Hence $f|D$ analytic implies $\int_{\partial D} f dz = 0$.

We use this to prove a version of Cauchy's Integral Formula:

5.8. Let $f : \mathbf{C} \to \mathbf{C}$ be differentiable, D the closed unit circle, $\zeta \in \mathring{D}$, then

$$f(\zeta) = \frac{1}{2\pi i} \int_{\partial D} \frac{f(z)}{z-\zeta} dz + \frac{1}{2\pi i} \int_D \frac{\partial f/\partial \bar{z}}{(z-\zeta)} dz \wedge d\bar{z} .$$

Proof. Let D_ε be a circle around ζ with radius ε, contained in $\mathring{D}$. The function $1/(z-\zeta)$ is analytic outside D_ε so applying Stokes' theorem to the function $f(z)/(z-\zeta)$ gives

$$\int_{D-D_\varepsilon} \frac{\partial f}{\partial \bar{z}} (z-\zeta)^{-1} dz \wedge d\bar{z} = \int_{\partial D} f(z)(z-\zeta)^{-1} dz - \int_0^{2\pi} f(\zeta+\varepsilon e^{i\theta}) i d\theta,$$

($\partial/\partial z$ and $\partial/\partial \bar{z}$ satisfy the product rule, as they should do).

As $\varepsilon \to 0$, the last integral converges to $2\pi i f(\zeta)$ and what is missing on the left-hand side, namely $\int_{D_\varepsilon} \frac{\partial f}{\partial \bar{z}} \cdot (z-\zeta)^{-1} dz \wedge d\bar{z}$, converges to zero, since $dz \wedge d\bar{z} = -2i dx \wedge dy = -2i r dr d\theta$ and $|z-\zeta|^{-1} = r^{-1}$. Hence $|\int_{D_\varepsilon} \dots| \le c \int_0^\varepsilon dr$, with $|2i \int_0^{2\pi} \frac{\partial f}{\partial \bar{z}} (\zeta + re^{i\theta}) d\theta| < c$. ✓

It is our aim to make statements of the following kind: Let $\tilde{f} : (\mathbf{R}^{n+1}, 0) \to \mathbf{R}$ be a germ with non-vanishing jet, then one may choose coordinates $(t, x_1, \dots, x_n)$ on $\mathbf{R}^{n+1}$ such that

$$\tilde{f}(t, x) = \tilde{Q}(t, x) \cdot \tilde{P}(t, x)$$

with

$$\tilde{Q}(0, 0) \neq 0, \text{ i. e. } \tilde{Q} \text{ a unit in } \mathcal{E}(n+1), \text{ and}$$

$$\tilde{P}(t, x) = t^p + \sum_{j=1}^{p} \tilde{\lambda}_j(x) t^{p-j}, \quad \tilde{\lambda}_j(0) = 0,$$

hence $\tilde{P} \in \mathcal{E}(n)[t]$.

Thus if a germ in $\mathcal{E}(n+1)$ has a non-vanishing Taylor series then, in suitable coordinates, it can be written as a polynomial over $\mathcal{E}(n)$ up to multiplication by a unit (a normed polynomial $P \in \mathcal{E}(n)[t]$ whose coefficients λ_j lie in $\mathfrak{m}(n)$ is called <u>distinguished</u>). In the analytic case, this fact may be used to deduce statements about $\mathcal{E}(n)$ inductively from statements about polynomials.

The proof of the above claim - which will be formulated later as a theorem - starts by showing that an arbitrary germ $\tilde{f} \in \mathcal{E}(n+1)$ can be <u>divided</u> by an arbitrary distinguished polynomial $\tilde{P} \in \mathcal{E}(n)[t]$ with a <u>remainder</u> of the form

$$\tilde{R}(t, x) = \sum_{j=1}^{p} \tilde{h}_j(x) t^{p-j},$$

that is, $\tilde{f} = \tilde{Q} . \tilde{P} + \tilde{R}$.

Next one shows that there are suitable coordinates and a suitable polynomial $\tilde{P}$ such that $\tilde{Q}$ is a unit and $\tilde{R} \equiv 0$. To simplify the proof, one introduces coefficients λ_j, $j = 1, \dots, p$, for a general polynomial P. At first these are regarded as new, additional coordinates which are independent of x. The λ_j are permitted to be complex and f does not depend on them. After completing the division, the λ_j are converted to functions $\lambda_j(x)$ by substitution of suitable germs from $\mathcal{E}(n)$. Hence we have to prove the following theorem.

5.9. Special Division Lemma. Let $\tilde{f} : (\mathbf{R} \times \mathbf{R}^n, 0) \to \mathbf{C}$ be a differentiable germ, and let $\tilde{P} : (\mathbf{R} \times \mathbf{C}^p, 0) \to \mathbf{C}$ be the germ of the 'general' polynomial

$$P(t, \lambda) = t^p + \sum_{j=1}^{p} \lambda_j t^{p-j},$$

then there exist differentiable germs $\tilde{Q}, \tilde{R} : (\mathbf{R}^{n+1} \times \mathbf{C}^p, 0) \to \mathbf{C}$, where

$$\tilde{R}(t, x, \lambda) = \sum_{j=1}^{p} \tilde{h}_j(x, \lambda).\, t^{p-j},$$

such that one has the following division with remainder:

$$\tilde{f}(t, x) = \tilde{Q}(t, x, \lambda).\, \tilde{P}(t, \lambda) + \tilde{R}(t, x, \lambda).$$

If $\tilde{f}, \lambda$ are real, then $\tilde{Q}, \tilde{R}$ can be chosen real.

For the proof, consider the classical case first: for fixed x, the function $f(t, x)$ is analytic in $t \in \mathbf{C}$ (as usual f is a representative of $\tilde{f}$).

Cauchy's integral formula gives:

$$f(t, x) = \frac{1}{2\pi i} \int_{\partial D} \frac{f(z, x)}{z - t}\, dz. \tag{1}$$

Here, and in what follows, D denotes a circular disc which contains all small t and all roots of $P(t, \lambda)$ for small λ. This exists by the earlier lemma (5.1).

The polynomial (with indeterminates z, t):

$$\frac{P(z, \lambda) - P(t, \lambda)}{z - t} = r(z, t, \lambda)$$

is analytic and, as a polynomial in t, has degree $< p$.

This equation yields the identity between rational functions:

$$\frac{1}{z - t} = \frac{P(t, \lambda)}{(z - t).\, P(z, \lambda)} + \frac{r(t, z, \lambda)}{P(z, \lambda)}. \tag{2}$$

Substitute (2) in (1) (the denominator does not vanish on ∂D if t and λ are sufficiently small). This gives the holomorphic (analytic) division lemma:

$$f(t, x) = P(t, \lambda) \cdot \underbrace{\frac{1}{2\pi i}\int_{\partial D} \frac{f(z, x)}{(z-t)P(z, \lambda)} dz}_{Q(t, x, \lambda)} + \underbrace{\frac{1}{2\pi i}\int_{\partial D} \frac{f(z, x)}{P(z, \lambda)} \cdot r(t, z, \lambda) dz.}_{R(t, x, \lambda)}$$

R is a polynomial in t of degree $< p$ (the integrand is a polynomial in t and the coefficients are integrated with respect to z). This completes the holomorphic division lemma.

To prove the corresponding result in the differentiable case, we use the version of Cauchy's integral formula given above and the following:

5.10. **Extension Lemma.** Let $f : \mathbf{R} \times \mathbf{R}^n \to \mathbf{C}$ be differentiable (with support in the unit ball), then there is a differentiable map

$$F : \mathbf{C} \times \mathbf{R}^n \times \mathbf{C}^p \to \mathbf{C},$$

such that (I) $F(t, x, \lambda) = f(t, x)$ for $(t, x) \in \mathbf{R}^{n+1} \subset \mathbf{C} \times \mathbf{R}^n$

(II) $\partial F/\partial \bar{z}$ vanishes to infinite order on $\{(z, x, \lambda) | \operatorname{Im} z = 0\}$ and on $\{(z, x, \lambda) | P(z, \lambda) = 0\}$.

The last condition states that the Taylor expansion of the map $\partial F/\partial \bar{z}$ vanishes at these points.

Assuming that the extension lemma is true we may continue the proof of the division lemma. From Cauchy's integral formula

$$f(t, x) = F(t, x, \lambda) = \frac{1}{2\pi i}\int_{\partial D} \frac{F(z, x, \lambda)}{z - t} dz + \frac{1}{2\pi i}\int_{D} \frac{F_{\bar{z}}(z, x, \lambda)}{z - t} dz \wedge d\bar{z}$$

where $F_{\bar{z}} = \partial F/\partial \bar{z}$, and D is as above. Substitute for $1/(z - t)$ from (2) to obtain $\tilde{f} = \tilde{Q}.\tilde{P} + \tilde{R}$ where

$$Q(t, x, \lambda) = \frac{1}{2\pi i}\int_{\partial D} \frac{F(z, x, \lambda)}{(z-t)P(z, \lambda)} dz + \frac{1}{2\pi i}\int_{D} \frac{F_{\bar{z}}(z, x, \lambda)}{(z-t)P(z, \lambda)} dz \wedge d\bar{z}$$

$$R(t, x, \lambda) = \frac{1}{2\pi i}\int_{\partial D} F(z, x, \lambda) \frac{r(t, z, \lambda)}{P(z, \lambda)} dz + \frac{1}{2\pi i}\int_{D} F_{\bar{z}}(z, x, \lambda) \frac{r(t, z, \lambda)}{P(z, \lambda)} dz \wedge d\bar{z}$$

The denominators do not vanish on ∂D, and we now have to show that the second integrals define differentiable maps. It is sufficient to show that the function

$$g(z, t, x, \lambda) = \frac{F_{\bar{z}}(z, x, \lambda)}{(z-t).\, P(z, \lambda)}\ ; \quad g = 0 \ \text{ for } \ z = t \ \text{ or } \ P(z, \lambda) = 0,$$

is differentiable. Each partial derivative of g is, <u>at a point where the denominator does not vanish</u>, a sum of functions of the form $F_0(z, x, \lambda)/[(z-t)P(z, \lambda)]^k$ where F_0 vanishes to infinite order on $\{\operatorname{Im} z = 0\}$ and $\{P(z, \lambda) = 0\}$. (Use the fact that the denominator is complex analytic in all variables, so that

$$\frac{\partial}{\partial \operatorname{Re} z}\left(\frac{1}{(z-t)P(z, \lambda)}\right) = \frac{\partial}{\partial z}\left(\frac{1}{(z-t)P(z, \lambda)}\right),$$

and similarly for the other variables. One can apply the quotient rule formally.) Because F_0 vanishes to arbitrarily high order on $\{\operatorname{Im} z = 0\}$, we have $F_0(z, x, \lambda) = (\operatorname{Im} z)^l F_1(z, x, \lambda)$, l arbitrarily large. F_1 vanishes to infinite order on $\{\operatorname{Re} P(z, \lambda) = \operatorname{Im} P(z, \lambda) = 0\}$ and may therefore be written:

$$F_1 = (\operatorname{Re} P(z, \lambda))^l F_2(z, x, \lambda) + (\operatorname{Im} P(z, \lambda))^l F_3(z, x, \lambda),$$

l arbitrarily large. (Observe that $(\operatorname{Re} P, \operatorname{Im} P)$ may be introduced as local coordinates, see 5.4.)

Now as (z, t, x, λ) approaches a point where $P.(z - t) = 0$ it is clear that

$$\frac{F_2 \,.\, (\operatorname{Re} P)^l \,.\, (\operatorname{Im} z)^l}{((z - t) \,.\, P)^k} \to 0, \text{ for } l > k.$$

The same is true of the second summand

$$\frac{F_3 \,.\, (\operatorname{Im} P)^l \,.\, (\operatorname{Im} z)^l}{((z - t) \,.\, P)^k}.$$

Hence all the derivatives of g converge to zero when the denominator tends to zero. It follows that g is differentiable, with derivative 0 where the denominator vanishes (see the problem at the end of this chapter). This establishes that Q and R are differentiable. If λ is real and f real-valued then one takes the real part of the equation

$$f(t, x) = Q(t, x, \lambda) \,.\, P(t, \lambda) + R(t, x, \lambda)$$

to get the real division

$$f = \tfrac{1}{2}(Q + \overline{Q})P + \tfrac{1}{2}(R + \overline{R}) .$$

It remains to prove the extension lemma. The proof is similar to that of Borel's theorem and is divided into three parts.

5.11. **Lemma.** Let $\mathbf{R} \subset \mathbf{C}$ be the standard imbedding, and $f : \mathbf{R} \times \mathbf{R}^n \to \mathbf{C}$ be differentiable (with support inside the unit ball), then there is a differentiable function $F : \mathbf{C} \times \mathbf{R}^n \to \mathbf{C}$, such that $F|\mathbf{R} \times \mathbf{R}^n = f$, and such that $\partial F/\partial \bar{z} : \mathbf{C} \times \mathbf{R}^n \to \mathbf{C}$ vanishes to infinite order on $\mathbf{R} \times \mathbf{R}^n$.

Proof. Let $z = x + iy$ and set

$$F(z) = \sum_{j=0}^{\infty} \left(i\frac{\partial}{\partial x}\right)^j f(x) \cdot \frac{y^j}{j!} \cdot \phi(t_j \cdot y),$$

where $\phi(y) = 1$ for $|y| \leq \frac{1}{2}$, $\phi(y) = 0$ for $|y| \geq 1$, and the sequence $\{t_j\}$ increases so rapidly that the series is differentiable term by term. Then $F(x) = f(x)$ for real x.

$$2\partial/\partial\bar{z} = \partial/\partial x + i\partial/\partial y = i(-i\partial/\partial x + \partial/\partial y),$$

and hence

$$\frac{2}{i}\frac{\partial}{\partial \bar{z}} F(z) = \sum_{j=0}^{\infty} \left(i\frac{\partial}{\partial x}\right)^{j+1} f(x) \cdot \frac{y^j}{j!} \cdot [\phi(t_{j+1} \cdot y) - \phi(t_j \cdot y)]$$
$$+ \sum_{j=0}^{\infty} \left(i\frac{\partial}{\partial x}\right)^j f(x) \cdot \frac{y^j}{j!} \cdot \phi'(t_j \cdot y),$$

and in both of these series each term vanishes on a neighbourhood of $y = 0$, because $\phi(t_j \cdot y)$ has locally the constant value 1. $\checkmark$

5.12. **Lemma.** Let $u, v : \mathbf{R}^m \times \mathbf{R}^n \times \mathbf{R}^k \to \mathbf{C}$ be differentiable functions such that $(u - v)$ vanishes to infinite order on $\{0\} \times \{0\} \times \mathbf{R}^k$. Then there exists $F : \mathbf{R}^m \times \mathbf{R}^n \times \mathbf{R}^k \to \mathbf{C}$ such that $(F - u)$ and $(F - v)$ vanish to infinite order on $\{0\} \times \mathbf{R}^n \times \mathbf{R}^k$ and $\mathbf{R}^m \times \{0\} \times \mathbf{R}^k$ respectively (the supports of u, v are contained in the unit ball).

Proof. Let $D^{\alpha,0,0}(u-v)(0, y, z) = f_\alpha(y, z)$ and

$$F(x, y, z) = v(x, y, z) + \sum_{(\alpha, 0, 0)} f_\alpha(y, z) \cdot \frac{x^\alpha}{\alpha!} \cdot \phi(t_{|\alpha|} \cdot x),$$

where $t_{|\alpha|}$ increases rapidly enough and $\phi = 1$ for $|x| \le \frac{1}{2}$, $\phi = 0$ for $|x| > 1$.

The series may be differentiated term by term. All derivatives of f_α vanish at $y = 0$ and so $D^\beta F(x, 0, z) = D^\beta v(x, 0, z)$. At $x = 0$, we have

$$\begin{aligned} D^{\alpha_1, \alpha_2, \alpha_3} F(0, y, z) &= D^{\alpha_1, \alpha_2, \alpha_3} v(0, y, z) + D^{\alpha_2, \alpha_3} f_{\alpha_1}(y, z) \\ &= D^{\alpha_1, \alpha_2, \alpha_3} v(0, y, z) + D^{\alpha_1, \alpha_2, \alpha_3}(u-v)(0, y, z) \\ &= D^{\alpha_1, \alpha_2, \alpha_3} u(0, y, z). \quad \checkmark \end{aligned}$$

Next we have the

Proof of the Extension Lemma. Let $f : \mathbf{R} \times \mathbf{R}^n \to \mathbf{C}$ be a given differentiable function (with compact support) and $P : \mathbf{C} \times \mathbf{C}^p \to \mathbf{C}$ a given polynomial with the form

$$P(z, \lambda) = z^p + \sum_j \lambda_j z^{p-j} .$$

We wish to extend f to $F : \mathbf{C} \times \mathbf{R}^n \times \mathbf{C}^p \to \mathbf{C}$ in such a way that $\partial F/\partial \bar{z}$ vanishes to infinite order on $\{\mathrm{Im}\, z = 0\}$ and $\{P = 0\}$.

<u>Induction on</u> p = Degree P

For $p = 0$, f must be extended so that $F_{\bar{z}}$ vanishes to infinite order on the real axis. This is possible by lemma 5.11.

Next, assume the extension lemma is true for $p - 1$. Change coordinates on $\mathbf{C} \times \mathbf{C}^p$ at 0 in the same way as before:

$$(z, \lambda_1, \ldots, \lambda_p) \mapsto (z, \lambda_1, \ldots, \lambda_{p-1}, P(z, \lambda)) = (z, \lambda', \mu), \text{ say.}$$

In the new coordinates, the operator $\partial/\partial\bar{z}$ becomes

$$L = \frac{\partial}{\partial \bar{z}} + \overline{P'(z, \lambda)} \cdot \frac{\partial}{\partial \bar{\mu}}$$

where P' is the derivative of P (P' is independent of λ_p).

Now we construct two differentiable, complex-valued functions

$v(z, x, \lambda')$, $u(z, x, \lambda', \mu)$, $z \in \mathbf{C}$, $\lambda' \in \mathbf{C}^{p-1}$, $\mu \in \mathbf{C}$, $x \in \mathbf{R}^n$,

with the following properties

(I) $v(t, x, \lambda') = f(t, x)$ for $t \in \mathbf{R}$

(II) $\partial v/\partial \bar{z}$ vanishes to infinite order on $\{\mathrm{Im}\, z = 0\}$

(III) $u = v$ for $\mu = 0$

(IV) $L(u)$ vanishes to infinite order for $\mu = 0$

(V) $(u - v)$ vanishes to infinite order for

$$\mathrm{Im}\, z = \mathrm{Re}\, \mu = \mathrm{Im}\, \mu = 0.$$

With this construction completed, lemma 5.12 supplies a differentiable function $F(z, x, \lambda', \mu)$ which satisfies: $(F - v)$ vanishes to infinite order on $\mathrm{Im}\, z = 0$, in particular $L(F)$ vanishes to infinite order and $F = v = f$ on $\mathrm{Im}\, z = 0$. This uses (II) and the fact that $\partial v/\partial \bar{\mu} \equiv 0$. Secondly $(F - u)$ vanishes to infinite order where $\mu = 0$. In particular, by IV, $L(F)$ vanishes to infinite order on $\mu = 0$.

<u>The construction</u>

Choose $v = v(z, x, \lambda')$ using the induction hypothesis, so that (I) and (II) are satisfied and $v_{\bar{z}}$ vanishes to infinite order when $P'(z, \lambda') = 0$ or $\mathrm{Im}\, z = 0$. Then put

$$u = \sum_{j=0}^{\infty} \left(-\frac{1}{\bar{P}'}\frac{\partial}{\partial \bar{z}}\right)^j v(z, \lambda') \cdot \frac{\bar{\mu}^j}{j!} \cdot \phi(t_j \cdot \mu)$$

where, by convention, $\left(-\frac{1}{\bar{P}'}\frac{\partial}{\partial \bar{z}}\right)^j v(z, \lambda')$ is zero when $P' = 0$ and $j \geq 1$. In this definition, ϕ is the function defined in 5.11 and t_j increases sufficiently quickly (the summands <u>are</u> differentiable). It follows that u is differentiable term by term and that (III) is satisfied.

$L(u)$ is computed term by term:

$$L(u) = \bar{P}' \left(\frac{1}{\bar{P}'}\frac{\partial}{\partial \bar{z}} + \frac{\partial}{\partial \bar{\mu}}\right)u$$

$$= -\bar{P}' \sum_{j=0}^{\infty} \left(-\frac{1}{\bar{P}'}\frac{\partial}{\partial \bar{z}}\right)^{j+1} v(z, \lambda') \cdot \frac{\bar{\mu}^j}{j!} \cdot [\phi(t_j \cdot \mu) - \phi(t_{j+1} \cdot \mu)]$$

$$+ \overline{P}' \sum_{j=0}^{\infty} \left(-\frac{1}{\overline{P}'} \frac{\partial}{\partial \overline{z}}\right)^j v(z, \lambda') \cdot \frac{\overline{\mu}^j}{j!} \cdot t_j \cdot \frac{\partial \phi}{\partial \overline{\mu}} (t_j \cdot \mu).$$

Now, observe that ϕ is locally constant around $\mu = 0$, and so all summands vanish there locally. This demonstrates (IV). To show (V), one checks that, because v satisfies (II), $u(z, x, \lambda', \mu) - v(z, x, \lambda') . \phi(t_0 . \mu)$ vanishes to infinite order on $\operatorname{Im} z = 0$. At a point where $\mu = 0$, evidently $v - v . \phi(t_0 . \mu)$ vanishes to infinite order. This completes the proof of the special division lemma. √

5.13. **Exercise.** Let $A \subset \mathbf{R}^n$ be closed and $f : \mathbf{R}^n \to \mathbf{R}$ a map with the following properties:

(I) $f | A = 0$

(II) $f | \mathbf{R}^n - A$ is differentiable

(III) If $a \in A$ is a boundary point, then for all

$$\alpha = (\alpha_1, \ldots, \alpha_n)$$

$$\lim_{\substack{x \to a \\ x \notin A}} D^{\alpha} f(x) = 0 .$$

Prove that f is differentiable.

6 · The preparation theorem

The literature is the same as for chapter 5.

6.1. Definition. A differentiable germ $\tilde{f} : (\mathbf{R} \times \mathbf{R}^n, 0) \to \mathbf{R} : (t, x) \mapsto f(t, x)$ is called p-regular (with respect to t), if $\tilde{f} | \mathbf{R} \times \{0\} \in m(1)^p$ and $\notin m(1)^{p+1}$. In other words

$$\tilde{f}(0, 0) = \frac{\partial}{\partial t}\tilde{f}(0, 0) = \ldots = \frac{\partial^{p-1}}{\partial t^{p-1}}\tilde{f}(0, 0) = 0, \quad \frac{\partial^p}{\partial t^p}\tilde{f}(0, 0) \neq 0.$$

For the jet of $\tilde{f}$, this means $\hat{f}(t, 0) = at^p$ + higher order terms, $a \neq 0$.

6.2. Remark. If $\tilde{f} \in \mathcal{E}(n+1)$ and $\hat{f} = j(f) \neq 0$, then there exists a linear isomorphism h of $\mathbf{R}^{n+1}$ and an integer p such that $\tilde{f} \circ \tilde{h}$ is p-regular. For p one may choose the smallest number such that $\tilde{f} \in m(n+1)^p$, $\tilde{f} \notin m(n+1)^{p+1}$.

Proof. $\hat{f} = \phi(x_1, \ldots, x_{n+1}) + \psi(x_1, \ldots, x_{n+1})$, where $\phi \neq 0$ is a homogeneous polynomial of degree p and $\psi \in \hat{m}(n+1)^{p+1}$. Choose an $a = (a_1, \ldots, a_{n+1}) \neq 0$ such that $\phi(a) \neq 0$. Choose a linear isomorphism $h : \mathbf{R}^{n+1} \to \mathbf{R}^{n+1}$ such that $h(1, 0, \ldots, 0) = a$, then

$$\begin{aligned}\hat{f} \circ \hat{h}(t, 0, \ldots, 0) &= \hat{f}(ta_1, \ldots, ta_{n+1}) \\ &= t^p . \underbrace{\phi(a_1, \ldots, a_{n+1})}_{\neq 0} + \underbrace{\psi(ta)}_{\in m(1)^{p+1}}. \quad \checkmark\end{aligned}$$

Next we have the result which we stated in the last chapter:

6.3. Division Theorem (Malgrange). Let $\tilde{f} : (\mathbf{R} \times \mathbf{R}^n, 0) \to \mathbf{R}$ be a p-regular germ with respect to the first variable, then there exist germs $\tilde{u}_1, \ldots, \tilde{u}_p \in m(n)$ and a unit $\tilde{Q} \in \mathcal{E}(n+1)$ such that

$$\tilde{f} = \tilde{Q}.\tilde{P}_u, \quad \tilde{P}_u(t, x) = t^p + \sum_{j=1}^{p} \tilde{u}_j(x)t^{p-j}.$$

By the previous remark, this means that a germ in $\mathscr{E}(n+1)$ with non-vanishing jet can, in a suitable coordinate frame, be written as a distinguished polynomial with coefficients in $\mathscr{E}(n)$, up to multiplication by a unit. Here $\mathscr{E}(n) \subset \mathscr{E}(n+1)$ is the ring of those germs which do not depend on the first variable.

6.4. Consequence: Generalised division lemma. Let $\tilde{f}, \tilde{g} \in \mathscr{E}(n+1)$ be germs where $\tilde{f}$ is p-regular, then there exists a $\tilde{Q} \in \mathscr{E}(n+1)$, and germs $\tilde{h}_j \in \mathscr{E}(n)$, $j = 1, \ldots, p$, such that

$$\tilde{g} = \tilde{Q}.\tilde{f} + \tilde{R}_h, \quad \tilde{R}_h(t, x) = \sum_{j=1}^{p} \tilde{h}_j(x)t^{p-j}.$$

Thus, instead of dividing by a polynomial, one may divide by an arbitrary p-regular germ in such a way that the remainder is a polynomial of degree $< p$ with coefficients in $\mathscr{E}(n)$.

Proof of the consequence. By the division theorem $\tilde{f} = \tilde{Q}_1.\tilde{P}_u$ for a unit $\tilde{Q}_1 \in \mathscr{E}(n+1)$ and a distinguished polynomial $P_u \in \mathscr{E}(n)[t]$. By the special division lemma

$$\tilde{g} = \tilde{Q}_2.\tilde{P}_u + \tilde{R}_h, \quad \tilde{R}_h(t, x) = \sum_{j=1}^{p} \tilde{h}_j(x)t^{p-j}$$

(here $\lambda_j = \tilde{u}_j(x)$ has been substituted in the special division lemma). Hence

$$\tilde{g} = (\tilde{Q}_2/\tilde{Q}_1).\tilde{f} + \tilde{R}_h. \quad \checkmark$$

Proof of the division theorem. We make the coefficients $\lambda = (\lambda_1, \ldots, \lambda_p) \in \mathbf{R}^p$ of the general polynomial into new variables again. By the special division lemma (division with remainder) we have:

$$\text{(1)} \quad \tilde{f}(t, x) = \tilde{Q}_1(t, x, \lambda).\tilde{P}(t, \lambda) + \tilde{R}(t, x, \lambda)$$

$$\tilde{P}(t, \lambda) = \sum_{j=0}^{p} \lambda_j t^{p-j}, \text{ where } \lambda_0 = 1$$

$$\tilde{R}(t, x, \lambda) = \sum_{j=1}^{p} \tilde{h}_j(x, \lambda)t^{p-j}.$$

The problem is to substitute germs $\tilde{u}_j(x)$ for the λ_j such that $\tilde{h}_j(x, \tilde{u}_j(x)) \in \mathcal{E}(n)$ vanishes. Now f is assumed p-regular with respect to t, and hence

(2) $\tilde{Q}_1(0, 0, 0) \neq 0, \quad \tilde{h}_j(0, 0) = 0$

(3) $\frac{\partial h_j}{\partial \lambda_i}(0, 0) = 0$ for $i < j$

(4) $\frac{\partial h_j}{\partial \lambda_j}(0, 0) \neq 0$

Proof of (2). From (1)

$$\tilde{f}(t, 0) = \tilde{Q}(t, 0, 0). t^p + \sum_{j=1}^{p} \tilde{h}_j(0, 0)t^{p-j}$$

for $x = \lambda = 0$. Hence, since this function vanishes to exactly order p, we must have (2).

Proof of (3), (4). Differentiate (1) at the point $x = \lambda = 0$ with respect to λ_i, to obtain

$$0 = t^{p-i}. \tilde{Q}_1(t, 0, 0) + t^p . \frac{\partial \tilde{Q}_1}{\partial \lambda_i}(t, 0, 0) + \sum_{j=1}^{p} \frac{\partial \tilde{h}_j}{\partial \lambda_i}(0, 0)t^{p-j}.$$

Modulo t^p this has the form

(5) $0 = t^{p-i}. q(t) + \sum_{j=1}^{p} h_{ji}t^{p-j}; \quad q(0) \neq 0$ by (2).

This is an equation of germs in one variable, t, considered in $\mathcal{E}(1)/m(1)^p \cong \mathbf{R}[t]/(t^p)$.

For $i < j$

Consider the last equation modulo t^{p-i} for fixed i, then

$$0 = \sum_{j=i+1}^{p} h_{ji}t^{p-j}, \text{ that is } h_{ji} = \frac{\partial h_j}{\partial \lambda_i}(0, 0) = 0.$$

For $i = j$

The equation (5) modulo t^{p-i+1} gives

$$0 = t^{p-i} \cdot q(0) + h_{ii} t^{p-i}, \quad \text{so that } h_{ii} \neq 0.$$

Now we come to the proof of the division theorem. Equation (2) shows that $\tilde{Q}_1 \in \mathcal{E}(n+1)$ is a unit. Further, the matrix $(h_{ji}) = (\partial h_j / \partial \lambda_i (0, 0))$ has triangular form with non-zero elements on the diagonal. Hence the equation $\tilde{h}(x, \lambda) = 0$ can be solved for the λ_j where

$$\tilde{h} : (\mathbf{R}^n \times \mathbf{R}^p, 0) \to \mathbf{R}^p$$

$$(x, \lambda) \mapsto (\tilde{h}_1(x, \lambda), \ldots, \tilde{h}_p(x, \lambda)).$$

This means that there exists a germ

$$\tilde{u} = (u_1, \ldots, u_p) : (\mathbf{R}^n, 0) \to (\mathbf{R}^p, 0)$$

such that $\tilde{h}(x, \tilde{u}(x)) = 0$. This is an application of the inverse function theorem that we have used before. The germ

$$\phi : (\mathbf{R}^n \times \mathbf{R}^p, 0) \to (\mathbf{R}^n \times \mathbf{R}^p, 0)$$

$$(x, \lambda) \mapsto (x, h(x, \lambda))$$

is invertible since its Jacobian matrix at the origin has the form

$$\left[\begin{array}{ccc|c} 1 & & 0 & \\ & \ddots & & 0 \\ 0 & & 1 & \\ \hline & ? & & h_{ji} \end{array}\right],$$

and u is the composite

$$\begin{array}{ccc} (\mathbf{R}^n, 0) = (\mathbf{R}^n \times \{0\}, 0) & \subset & (\mathbf{R}^n \times \mathbf{R}^p, 0) \\ \quad\searrow u & & \downarrow \phi^{-1} \\ (\mathbf{R}^p, 0) & \xleftarrow{\text{proj}_2} & (\mathbf{R}^n \times \mathbf{R}^p, 0) \end{array}$$

If the germ $\tilde{u}$ is substituted for λ in (1), then equation (2) and $\tilde{h}(x, \tilde{u}(x)) = 0$ give the division theorem. √

This brings us to one of the major results in this book:

6.5. **The preparation theorem of Malgrange** (in the form from J. Mather). Let $f : (\mathbf{R}^n, 0) \to (\mathbf{R}^p, 0)$ be a differentiable germ, it induces the homomorphism $f^* : \mathcal{E}(p) \to \mathcal{E}(n)$ of rings. Let A be a finitely generated $\mathcal{E}(n)$-module, then: A is finitely generated over $\mathcal{E}(p)$ (operating on A via f^*) if and only if the real vector space $A/(f^*\mathfrak{m}(p).A)$ is finite dimensional.

Proof. Observe that $(f^*\mathfrak{m}(p).A)$ is the same as $\mathfrak{m}(p).A$ where the operation is via f^*.

One direction is trivial: If A is finitely generated over $\mathcal{E}(p)$, then one has an epimorphism of $\mathcal{E}(p)$-modules

$$\bigoplus_{j=1}^{k} \mathcal{E}(p) = \underbrace{\mathcal{E}(p) \oplus \ldots \oplus \mathcal{E}(p)}_{k \text{ summands}} \to A$$

and therefore an epimorphism (modulo multiples of $\mathfrak{m}(p)$)

$$\mathbf{R}^k = \bigoplus_{j=1}^{k} \mathcal{E}(p)/\mathfrak{m}(p) \to A/(f^*\mathfrak{m}(p).A) .$$

In other words, the generators of A over $\mathcal{E}(p)$ are also generators of $A/(f^*\mathfrak{m}(p).A)$ over $\mathcal{E}(p)/\mathfrak{m}(p) = \mathbf{R}$.

The other direction is the deep theorem. Nevertheless we have already done all the work and the three steps that remain are pure pleasure.

Step 1 Let $n = p + 1$ and $\tilde{f} : (\mathbf{R} \times \mathbf{R}^p, 0) \to (\mathbf{R}^p, 0)$

$$(t, x) \mapsto x$$

be the projection onto the second factor (special case). In this case choose $a_1, \ldots, a_k \in A$, finitely many elements which generate both A as an $\mathcal{E}(p+1)$-module and $A/(f^*\mathfrak{m}(p).A)$ as a real vector space. Then any $a \in A$ may be written

$$a = \sum_{j=1}^{k} c_j a_j + \sum_{j=1}^{k} z_j a_j$$

$$c_j \in \mathbf{R},\ z_j \in f^*\mathfrak{m}(p).\mathscr{E}(p+1),$$

as follows: $a = \Sigma c_j a_j + b$, $b \in (f^*\mathfrak{m}(p).A)$, hence $b = \Sigma y_l b_l$, $y_l \in f^*\mathfrak{m}(p)$. Next $b_l = \Sigma r_{lj} a_j$, $r_{lj} \in \mathscr{E}(p+1)$, so write $z_j = \Sigma y_l r_{lj}$.

For $a = ta_i$ this gives, in particular,

$$ta_i = \sum_{j=1}^{k} (c_{ij} + z_{ij}) a_j,$$

$$c_{ij} \in \mathbf{R};\ z_{ij} \in f^*\mathfrak{m}(p).\mathscr{E}(p+1).$$

If (δ_{ij}) is the unit matrix, then this equation may be written

$$(t\delta_{ij} - c_{ij} - z_{ij}).\mathbf{a} = 0, \text{ where } \mathbf{a} = (a_1, \ldots, a_k).$$

Let $b_{ij} = t\delta_{ij} - c_{ij} - z_{ij}$. Using linear algebra we take a matrix (B_{ij}) (the transpose of the matrix of cofactors of the b_{ij}) such that

$$(B_{ij}).(b_{ij}) = \det(b_{ij}).(\delta_{ij}).$$

Put $\Delta(t, x) = \det(t\delta_{ij} - c_{ij} - z_{ij})$, then it follows that $\Delta.\mathbf{a} = 0$. This determinant is a function of $(t, x) \in \mathbf{R} \times \mathbf{R}^p$ and when $x = 0$ it is a normed polynomial in t (we have $z_{ij}(t, 0) = 0$ so Δ is the characteristic polynomial of (c_{ij}) at $x = 0$). We deduce that Δ is q-regular with respect to t at $(t, 0)$ for some $q \leq k$.

Since $\Delta\mathbf{a} = 0$, it follows that $\Delta A = 0$ and so A is a module over $\mathscr{E}(p+1)/\Delta.\mathscr{E}(p+1)$. Because Δ is q-regular, the generalised division lemma implies that the $\mathscr{E}(p)$-module $\mathscr{E}(p+1)/\Delta.\mathscr{E}(p+1)$ is generated by finitely many elements, namely $1, t, \ldots, t^{q-1}$. Now since A is finitely generated over $\mathscr{E}(p+1)/\Delta.\mathscr{E}(p+1)$ which is in turn finitely generated over $\mathscr{E}(p)$, we find that A is a finitely generated $\mathscr{E}(p)$-module.

Step 2 Let $\tilde{f} : (\mathbf{R}^n, 0) \to (\mathbf{R}^p, 0)$ be a germ with rank n. By the rank theorem, there are coordinates giving $\tilde{f}$ the form $(x_1, \ldots, x_n) \mapsto (x_1, \ldots, x_n, 0, \ldots, 0)$. Now for a canonical imbedding $\mathbf{R}^n \subset \mathbf{R}^p$, any differentiable germ $\tilde{\phi} : (\mathbf{R}^n, 0) \to \mathbf{R}$ may be extended to $(\mathbf{R}^p, 0)$. So the map $f^* : \mathscr{E}(p) \to \mathscr{E}(n)$ is surjective for this case. This means that a finite number of generators of A as an $\mathscr{E}(n)$-module are also generators of A as an $\mathscr{E}(p)$-module.

Next write an arbitrary germ $\tilde{f} : (\mathbf{R}^n, 0) \to (\mathbf{R}^p, 0)$ as the composite

$$(\mathbf{R}^n, 0) \xrightarrow[(\mathrm{id}, \tilde{f})]{} (\mathbf{R}^n \times \mathbf{R}^p, 0) \xrightarrow[\mathrm{pr}_2]{} (\mathbf{R}^p, 0) .$$

The first germ is an immersion and the second a sequence of n projections of the kind in step 1. Let $M(\tilde{f})$ be used to denote the property that: $A/(f^* \mathfrak{m}(p). A)$ finite dimensional $\Rightarrow$ A finitely generated over $\mathscr{E}(p)$. Then what we have to prove is:

Step 3 If

$$(\mathbf{R}^n, 0) \xrightarrow{\tilde{f}} (\mathbf{R}^p, 0) \xrightarrow{\tilde{g}} (\mathbf{R}^q, 0)$$

are differentiable germs then $M(\tilde{f})$ and $M(\tilde{g})$ imply $M(\tilde{g} \circ \tilde{f})$. Thus we suppose that A is a finitely generated $\mathscr{E}(n)$-module and

$$A/(\tilde{g} \circ \tilde{f})^* \mathfrak{m}(q). A = A/f^*(g^* \mathfrak{m}(q)). A$$

is finite dimensional over $\mathbf{R}$. Since $g^* \mathfrak{m}(q) \subset \mathfrak{m}(p)$, we have $f^* g^* \mathfrak{m}(q) \subset f^* \mathfrak{m}(p)$ and so $A/f^* \mathfrak{m}(p). A$ is finite dimensional. By $M(\tilde{f})$, it follows that A is finitely generated as an $\mathscr{E}(p)$-module via f^*.

Now, by definition, $A/g^* \mathfrak{m}(q). A = A/f^* g^* \mathfrak{m}(q). A$ and this has finite dimension. By $M(\tilde{g})$, it follows that the $\mathscr{E}(p)$-module A is finitely generated as an $\mathscr{E}(q)$-module via g^*, that is, the $\mathscr{E}(n)$-module A is finitely generated as an $\mathscr{E}(q)$-module via $(\tilde{g} \circ \tilde{f})^*$. ✓

The preparation theorem is now completely proved. There is the following slight extension of the result:

6.6. Preparation theorem - Corollary. With the hypotheses of the preparation theorem, the elements $\{a_1, \ldots, a_k\}$ generate A as an $\mathscr{E}(p)$-module if and only if they represent generators of the real vector space $A/(f^* \mathfrak{m}(p). A)$.

Proof. It has already been remarked that one direction is trivial. Let $\{a_1, \ldots, a_k\}$ be a system of generators of $A/(f^* \mathfrak{m}(p). A)$, so that from the preparation theorem A is finitely generated over $\mathscr{E}(p)$. Now

$$A = \langle a_1, \ldots, a_k \rangle_{\mathcal{E}(p)} + m(p).A,$$

where the first term on the right is the module generated over $\mathcal{E}(p)$ via f^* and the second is defined via f^*.

It follows from the Nakayama lemma that $A = \langle a_1, \ldots, a_k \rangle_{\mathcal{E}(p)}$. ✓

The special case $A = \mathcal{E}(n)$ is especially important:

6.7. **Preparation theorem** (in the Malgrange form). Let $\tilde{f} : (\mathbf{R}^n, 0) \to (\mathbf{R}^p, 0)$ be a differentiable germ. It induces the ring homomorphism $f^* : \mathcal{E}(p) \to \mathcal{E}(n)$, and the homomorphism $\hat{f}^* : \hat{\mathcal{E}}(p) \to \hat{\mathcal{E}}(n)$ of power series rings. The following are equivalent:

(I) $\phi_1, \ldots, \phi_k \in \mathcal{E}(n)$ generate $\mathcal{E}(n)$ as an $\mathcal{E}(p)$-module via f^*.

(II) $\hat{\phi}_1, \ldots, \hat{\phi}_k$ generate $\hat{\mathcal{E}}(n)$ as an $\hat{\mathcal{E}}(p)$-module via f^*.

(III) $\phi_1, \ldots, \phi_k$ represent generators of the real vector space $\mathcal{E}(n)/f^* m(p).\mathcal{E}(n)$.

(IV) $\hat{\phi}_1, \ldots, \hat{\phi}_k$ represent generators of the real vector space $\hat{\mathcal{E}}(n)/\hat{f}^* \hat{m}(p).\hat{\mathcal{E}}(n)$.

Proof. The equivalence of (I) and (III) is the preparation theorem in the extended form, with $\mathcal{E}(n) = A$.

(III) $\Rightarrow$ (IV). From $\phi_1.\mathbf{R} + \ldots + \phi_k.\mathbf{R} + f^* m(p).\mathcal{E}(n) = \mathcal{E}(n)$, it follows using the map $j : \mathcal{E}(n) \to \hat{\mathcal{E}}(n)$ that:

$$\hat{\phi}_1\mathbf{R} + \ldots + \hat{\phi}_k\mathbf{R} + \hat{f}^* \hat{m}(p).\hat{\mathcal{E}}(n) = \hat{\mathcal{E}}(n) .$$

(IV) $\Rightarrow$ (III). From (IV), $\mathcal{E}(n)/(m(p).\mathcal{E}(n) + m(n)^\infty)$ is finite dimensional. By Nakayama's lemma we deduce

$$m(n)^k/(\ldots) \underset{\neq}{\supset} m(n)^{k+1}/(\ldots) \text{ unless } m(n)^k/(\ldots) = 0$$

so that (for some k) $m(n)^k/(\ldots) = 0$. This means

$$m(n)^k \subset m(p).\mathcal{E}(n) + m(n)^\infty \subset m(p).\mathcal{E}(n) + m(n)^{k+1}.$$

By Nakayama's lemma $m(n)^k \subset m(p).\mathcal{E}(n)$, so that $\mathcal{E}(n)/m(p).\mathcal{E}(n)$ is equal to $\mathcal{E}(n)/(m(p).\mathcal{E}(n) + m(n)^k)$.

The last space is the image of $\mathcal{E}(n)/(m(p)\mathcal{E}(n) + m(n)^\infty)$ under projection and so generated over $\mathbf{R}$ by $\phi_1, \ldots, \phi_k$.

(I) $\Rightarrow$ (II). This follows trivially by going over to jets.

(II) $\Rightarrow$ (IV). This is simple and proceeds as in the corresponding part of the differentiable preparation theorem. ✓

The equivalence of (II) and (IV) is the formal preparation theorem and is a by-product of the proof for the real case. However, as with the formal inverse function theorem in chapter 4 ($D\hat{f}(0)$ invertible $\Longleftrightarrow$ $\hat{f}$ invertible), this result can be proved much more easily - and in a more general setting.

The ideal $m(p)$ is generated by the coordinate germs $(\tilde{y}_1, \ldots, \tilde{y}_p)$ on $\mathbf{R}^p$. Therefore if $\tilde{f} = (\tilde{f}_1, \ldots, \tilde{f}_p)$, we have $f^*\tilde{y}_j = \tilde{y}_j \circ \tilde{f} = \tilde{f}_j$ and so

$$\mathcal{E}(n).\, m(p) = \langle \tilde{f}_1, \ldots, \tilde{f}_p \rangle_{\mathcal{E}(n)}$$

is the ideal in the ring $\mathcal{E}(n)$, which is generated by the component functions of $\tilde{f}$.

6. 8. Definition. A differentiable germ $\tilde{f} : (\mathbf{R}^n, 0) \to (\mathbf{R}^p, 0)$ is called <u>finite</u>, if $\mathcal{E}(n)/\tilde{f}^* m(p). \mathcal{E}(n)$ has finite dimension.

6. 9. Exercise. If $\tilde{f} : (\mathbf{R}^n, 0) \to (\mathbf{R}^p, 0)$ is finite, prove there is a representative f such that each point in $\mathbf{R}^p$ has at most finitely many points in its inverse image. Hint: first obtain $f^{-1}(0)$ finite; then use the fact that a germ 'near' $\tilde{f}$ is also finite (see chapter 13).

6. 10. From the preparation theorem in Malgrange's form one may easily deduce the generalised division lemma. Let $F(t, x_1, \ldots, x_n)$ be p-regular with respect to t. Consider the germ $\tilde{f}(t, x_1, \ldots, x_n) = (\tilde{F}(t, x), x_1, \ldots, x_n)$. From the p-regularity of F it follows that

$$\langle F, x_1, \ldots, x_n \rangle_{\mathcal{E}(n+1)} = \langle t^p, x_1, \ldots, x_n \rangle_{\mathcal{E}(n+1)}.$$

By the preparation theorem $\mathcal{E}(n + 1)$ will be finitely generated as an $\mathcal{E}(n+1)$-module via f^* by $\{1, t, t^2, \ldots, t^{p-1}\}$. That is, for each $g \in \mathcal{E}(n + 1)$ one has

$$\tilde{g}(t, x_1, \ldots, x_n) = \sum_{i=1}^{p} \tilde{g}_i(F(t, x), x_1, \ldots, x_n) t^{p-i}$$

for certain $\tilde{g}_i \in \mathscr{E}(n+1)$. If we now put $\tilde{h}_i(x) = \tilde{g}_i(0, x)$, then $\tilde{g}_i(\tau, x) - \tilde{h}_i(x) = \tau . \tilde{k}_i(\tau, x)$ and so, substituting F for τ,

$$\tilde{g}(t, x_1, \ldots, x_n) = \sum_{i=1}^{p} \tilde{h}_i(x) t^{p-i} + \tilde{F}(t, x) . \tilde{Q}(t, x)$$

where $\tilde{Q}(t, x) = \sum_{i=1}^{p} \tilde{k}_i(F(t, x), x) t^{p-i}$.

For an $\mathscr{E}(n)$-module A it is easy to define an $\hat{\mathscr{E}}(n)$-module $\hat{A}$ and then possible to formulate a theorem in the style of Malgrange, even in the more general situation considered by Mather.

6.11. **Exercise.** Let $\tilde{f} : (\mathbf{R}^n, 0) \to (\mathbf{R}^p, 0)$ be a differentiable germ and $\dim(\mathscr{E}(n)/f^* \mathfrak{m}(p) . \mathscr{E}(n)) = k$. Prove that the $\mathscr{E}(p)$-module $\mathscr{E}(n)$ is generated by monomials of degree $< k$ in the coordinates on $\mathbf{R}^n$. (Hint: Nakayama.)

7 · Symmetric germs

This chapter is designed to show the preparation theorem in action in a simple case.

Recall that, in chapter 5, we defined the elementary symmetric functions $(-1)^i\sigma_i(x_1, \ldots, x_n)$, $i = 1, \ldots, n$, by

$$\prod_{i=1}^{n} (t - x_i) = \sum_{i=0}^{n} t^i \sigma_{n-i}(x); \quad \sigma_0 = 1.$$

In particular $\sum_{i=0}^{n} x_j^i \sigma_{n-i}(x) = 0$, or

$$x_j^n = - \sum_{i=1}^{n} x_j^i \sigma_{n-i}(x).$$

Denote by $\sigma : \mathbf{R}^n \to \mathbf{R}^n$ the map whose components are the σ_i for $i = 1, \ldots, n$. The above equation gives $x_j^n \in \sigma^* \mathfrak{m}(n). \mathcal{E}(n)$. Hence any monomial in the x_j, which contains an exponent $\geq n$, lies in $\sigma^* \mathfrak{m}(n). \mathcal{E}(n)$. It follows that

$$\mathfrak{m}(n)^k \subset \sigma^* \mathfrak{m}(n). \mathcal{E}(n) \text{ for } k \geq n^n.$$

Because the monomials of degree $< k$ generate the vector space $\mathcal{E}(n)/\mathfrak{m}(n)^k$, the preparation theorem gives us

7.1. **Lemma.** The monomials of degree $< n^n$ generate $\mathcal{E}(n)$ as a module over the ring of germs $\tilde{f}(\sigma_1, \ldots, \sigma_n)$. ✓

This has the following consequence:

7.2. **Definition.** A germ $\tilde{f} \in \mathcal{E}(n)$ is called symmetric, if for any permutation π of $\{1, \ldots, n\}$ the germ satisfies

$$\tilde{f}(x_1, \ldots, x_n) = \tilde{f}(x_{\pi(1)}, \ldots, x_{\pi(n)}).$$

7.3. Theorem (Glaeser). A symmetric differentiable germ may be written as a differentiable germ of the elementary symmetric functions.

Hence $\tilde{f}$ is symmetric if and only if there is a differentiable germ $\tilde{g} \in \mathscr{E}(n)$, with $\tilde{f} = \tilde{g} \circ \tilde{\sigma}$.

Proof. Let $\phi_1, \ldots, \phi_r$ be the monomials of degree $< n^n$ in the coordinate functions x_i. Let f be symmetric. By the lemma

$$f(x) = \sum_{j=1}^{r} \phi_j(x). g_j(\sigma(x)).$$

Let $\mathfrak{S}(n)$ be the group of permutations of $\{1, \ldots, n\}$, then, since f and the σ_i are symmetric,

$$f(x) = \frac{1}{n!} \sum_{j=1}^{r} \Big(\sum_{\pi \in \mathfrak{S}(n)} \phi_j(x_{\pi(1)}, \ldots, x_{\pi(n)})\Big) g_j(\sigma(x))$$

and the polynomials $p_j(x) = \sum_{\pi \in \mathfrak{S}(n)} \phi_j(x_{\pi(1)}, \ldots, x_{\pi(n)})$ are obviously symmetric. By the main theorem on symmetric polynomials (Lang, Algebra V, §9, p. 133) it follows that

$$p_j(x) = q_j(\sigma(x))$$

for a (unique) polynomial q_j.

Hence

$$\tilde{f}(x) = \frac{1}{n!} \sum_{j=1}^{r} q_j(\sigma(x)). g_j(\sigma(x)). \quad \checkmark$$

7.4. Exercises. 1. Let $f \in \mathscr{E}(n)$ and

$$f(x_1, \ldots, x_n) = f(x_1, \ldots, x_{i-1}, -x_i, x_{i+1}, \ldots, x_n)$$

for $1 \leq i \leq n$. Show that $f(x) = g(x_1^2, \ldots, x_n^2)$ for some $g \in \mathscr{E}(n)$.

2. Determine a basis of the ring of polynomials in n variables as a module over the ring of symmetric polynomials. (See Artin: Galois theory.)

8 · Mappings of the plane into the plane

Literature: H. Whitney: On singularities of mappings of Euclidean spaces I, mappings of the plane into the plane; Annals of Math., 62 (1955), 374-410.

B. Malgrange: Ideals of differentiable functions, Bombay lecture notes, Oxford Univ. Press (1966).

J. Milnor: Morse theory, Annals of Math. Studies 51 (1963).

R. Narasimhan: Analysis on real and complex manifolds, Masson and Cie, Paris, and North-Holland, Amsterdam (1968).

The set of differentiable maps $\mathbf{R}^n \to \mathbf{R}^p$ is denoted by $C^\infty(\mathbf{R}^n, \mathbf{R}^p)$ and made into a topological space by giving a base of neighbourhoods $\{U(\varepsilon, k)\}$ of the zero-function. For each $k \in \mathbf{N}$ and each continuous, strictly positive function $\varepsilon : \mathbf{R}^n \to \mathbf{R}$ define $U(\varepsilon, k)$ by

$$f \in U(\varepsilon, k) \iff |D^\alpha f_j(x)| < \varepsilon(x) \quad \text{for} \quad j = 1, \ldots, p,$$

$$\text{for all } |\alpha| \leq k, \ x \in \mathbf{R}^n.$$

Such a topology can also be put on the set $C^\infty(M, N)$, where M, N are differentiable manifolds, by imbedding M and N in Euclidean space. We shall not pursue the details (see Narasimhan).

If we say 'f is near to g' or 'f is small' then we shall be referring to this topology. If $K \subset \mathbf{R}^n$ is compact then the set $C^\infty(K, \mathbf{R}^p)$ has the quotient topology induced from $C^\infty(\mathbf{R}^n, \mathbf{R}^p)$ by the restriction map: $C^\infty(\mathbf{R}^n, \mathbf{R}^p) \to C^\infty(K, \mathbf{R}^p)$. This also defines the expression '$f : \mathbf{R}^n \to \mathbf{R}^p$ is small on K'.

We now begin the study of the following concepts.

8.1. Definition. Let M, N be differentiable manifolds. Two differentiable maps $f_0, f_1 : M \to N$ are called C^∞-equivalent (resp. topologically equivalent) if there are diffeomorphisms (resp. homeomorphisms) $h : M \to M$; $k : N \to N$ such that the following diagram commutes:

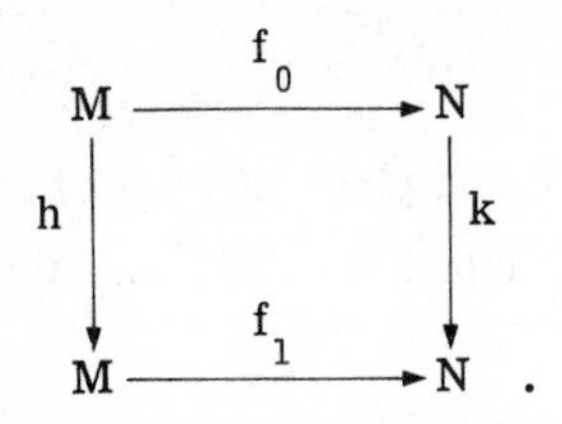

It is too much to ask for the equivalence classes of these relations to be completely described. For example, problem 2 in chapter 3 shows that any closed set in $\mathbf{R}^n$ is the set of singular points of some map $f : \mathbf{R}^n \to \mathbf{R}^n$.

The classification of differentiable maps, under the equivalence relations above, is thus immense, involving more than the classification of all closed subsets of $\mathbf{R}^n$.

The most one might hope for is that an open and dense subset of differentiable maps (in the topological sense 'almost every' differentiable map) can be described up to equivalence. Most of all one would like to have an open and dense subset of maps with the following property:

8.2. Definition. A differentiable map $f : M \to N$ is called C^∞-stable (resp. topologically stable), if there is a neighbourhood U of f in $C^\infty(M, N)$ such that each map $f_1 \in U$ is C^∞-equivalent (resp. topologically equivalent) to f.

The stable maps certainly form an open set in $C^\infty(M, N)$, but the questions are whether the set is dense and whether the stable maps can be classified. The second problem is very much at the bottom of things. We can define an equivalence relation on germs $(M, x) \to (N, f(x))$ similar to 8.1, and ask:

Are there finitely many germs: $(\mathbf{R}^m, 0) \to (\mathbf{R}^n, 0)$ such that if $f : M^m \to N^n$ is stable, then each germ $\tilde{f} : (M, x) \to (N, f(x))$ defined by f is equivalent to one of these finitely many?

One may suppose, that the natural geometric forms are described by stable maps (invariant under small disturbances). Hence a classification of germs of stable maps is at the same time a local classification of natural geometric forms.

The first result in this direction is given by Morse theory. Almost every differentiable function $f : M \to \mathbf{R}$ is stable, and each germ of a stable function is equivalent to one of the types

A) $(x_1, \ldots, x_n) \mapsto x_1$ (regular point)

B) $(x_1, \ldots, x_n) \mapsto x_1^2 + \ldots + x_k^2 - (x_{k+1}^2 + \ldots + x_n^2)$
(a critical point of index $n - k$)

This is easily obtained from the representation in Milnor's 'Morse Theory'.

The next result originates from Whitney and describes differentiable maps: $\mathbf{R}^2 \to \mathbf{R}^2$.

In particular, one finds that only three equivalence classes of germs arise. Imagine a flexible sheet being crumpled. The picture which is suggested to the mind is the typical image of a stable map from the plane into the plane. What does the map look like locally (in suitable coordinates)?

8.3. There are three possibilities:

(I) A neighbourhood of $x \in \mathbf{R}^2$ will be mapped regularly.

(II) The point x lies on a fold.

(III) The point x lies where a fold ends or begins.

For each of these cases it is simple to give an analytic example:

(I) $f : \mathbf{R}^2 \to \mathbf{R}^2$; $(x, z) \mapsto (x, z)$, regular point.

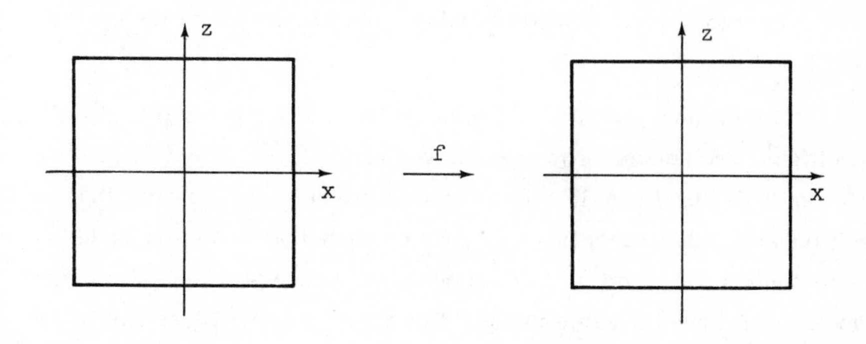

(II) $f : \mathbf{R}^2 \to \mathbf{R}^2$, $(x, z) \mapsto (x, z^2) = (x, y)$, fold.

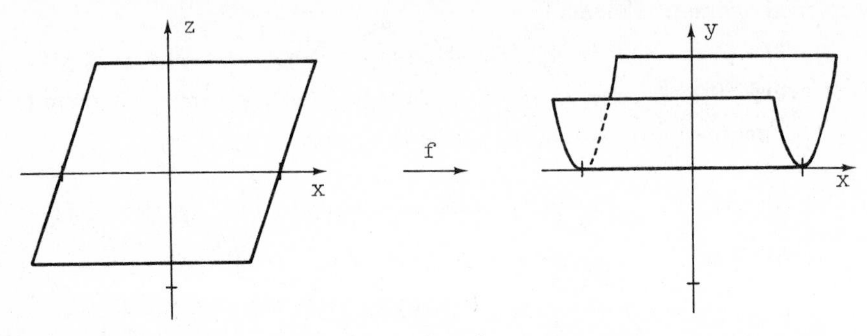

(III) $f : \mathbf{R}^2 \to \mathbf{R}^2$, $(x, z) \mapsto (x, z^3 - xz) = (x, y)$, cusp.

The second coordinate $y = z^3 - xz$, for fixed x, is a cubic with derivative $\partial y/\partial z = 3z^2 - x$. The extrema are given by $x = 3z^2$ and their values are $y = \pm(2/3\sqrt{3})x^{3/2}$. This gives the following picture.

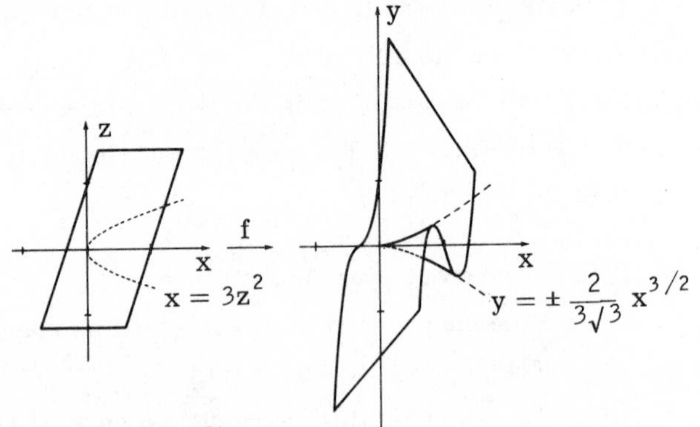

We wish to prove the following result:

8.4. **Theorem (H. Whitney).** There is an open and dense subset $T \subset C^\infty(\mathbf{R}^2, \mathbf{R}^2)$ such that for $x \in \mathbf{R}^2$ and $f \in T$, the germ $\tilde{f} : (\mathbf{R}^2, x) \to (\mathbf{R}^2, f(x))$ is differentiably equivalent to one of the germs (I), (II) or (III).

In fact those maps in T which satisfy a few other mild, global conditions, are stable. These conditions are: (1) the folds (which are differentiable curves in $\mathbf{R}^2$) should cut each other transversally (i. e. with linearly independent tangents), (2) no more than two folds should meet at any point, and (3) two cusp points or a cusp point and a fold point should never have the same image. The theorem generalises easily to maps from surfaces to surfaces (see picture)

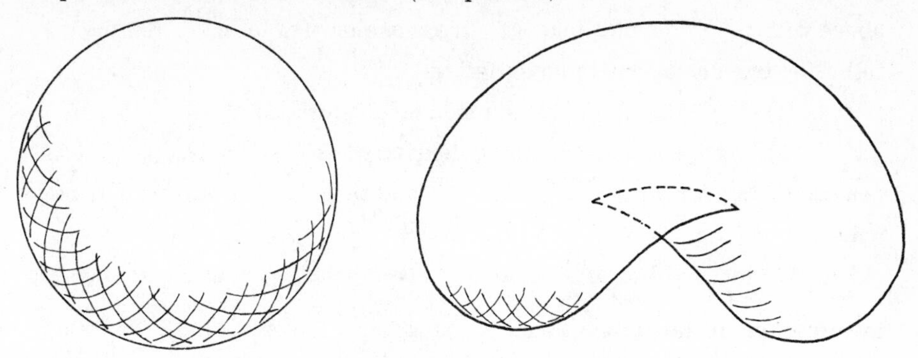

The proof of the theorem fills the rest of this chapter.

8.5. **Lemma.** Let $X = Y = \mathbf{R}^2$; $X = \{(x_1, x_2) | x_i \in \mathbf{R}\}$; $Y = \{(y_1, y_2) | y_j \in \mathbf{R}\}$, and let $f = (f_1, f_2) : X \to Y$ be differentiable. Then in each neighbourhood of f there is a differentiable map $g = (g_1, g_2)$, such that $\mathrm{Rk}_x g = \mathrm{Rk}(\partial g_i / \partial x_j)(x) \neq 0$.

Proof. Fix a neighbourhood of f. Consider the map

$$(\partial f_i / \partial x_j)_{i, j=1, 2} : \mathbf{R}^2 \to \mathbf{R}^4 .$$

This map is differentiable, and by Sard's theorem (trivial case), the image has measure zero. Arbitrarily near $0 \in \mathbf{R}^4$, there is a $\lambda = (\lambda_{ij})$ which is not in the image of the map. The map

$$\bar{g}_i(x_1, x_2) = f_i(x_1, x_2) - \lambda_{i1}x_1 - \lambda_{i2}x_2$$

has a non-vanishing derivative.

Next choose a differentiable function $\phi_n : \mathbf{R}^2 \to \mathbf{R}$ such that $\phi_n | K(0, n) = 1$ where $K(0, n)$ is the ball of radius n and centre at 0, and $\phi_n | \mathbf{R}^2 - K(0, n+1) = 0$. Put

$$g^1 = \phi_1 \bar{g} + (1 - \phi_1)f.$$

g^1 is near f for small λ and satisfies $(\partial g_i^1 / \partial x_j) \neq 0$ on $K(0, 1)$ for almost all λ. Construct g^n inductively putting

$$g^n = (\phi_n - \phi_{n-2})\bar{g} + (1 - (\phi_n - \phi_{n-2}))g^{n-1}, \quad \text{where } \phi_{-1} = 0.$$

g^n and g^{n-1} differ only on $K(0, n+1) - K(0, n-2)$; $\bar{g}$ is chosen as above with λ so small that g^n lies close enough to g^{n-1} for the following two conditions to be satisfied:

(1) g^n lies in the prescribed neighbourhood of f,

(2) g^n has non-vanishing derivative on $K(0, n-1)$, (g^{n-1} has non-vanishing derivative on $K(0, n-1)$ and therefore so will nearby maps, note that $g^n = \bar{g}$ on $K(0, n) - K(0, n-1)$).

We let $g = \lim_{n\to\infty} g^n$. Near any given point, g and g^n are locally identical for all large n. Hence g is differentiable and $(\partial g_i/\partial x_j) \neq 0$ everywhere. Also g lies in the given neighbourhood of f. ✓

8.6. The essential features of this construction, which constantly recurs in differential topology, are

(1) the property that one seeks to obtain (here, $Rk_x g \neq 0$ everywhere) defines an open subset of maps,

(2) each map may be approximated locally by maps with this property.

A germ $g : (\mathbf{R}^2, 0) \to (\mathbf{R}^2, 0)$, with non-vanishing derivative satisfies $\partial g_1/\partial x_1(0) \neq 0$ without loss of generality. The transformation $(x_1, x_2) \mapsto (g_1(x), x_2)$ in the domain gives the germ g the form $(x, z) \mapsto (x, \phi(x, z))$.

The preceding lemma shows that almost any map has the form just described, with respect to suitable coordinates. Further, if a map has this form on a compact neighbourhood then any nearby map can be put in this form on this neighbourhood.

The other changes needed for the main theorem may be made to the function $\phi : \mathbf{R}^2 \to \mathbf{R}$ just defined. We work with maps $\mathbf{R}^2 \to \mathbf{R}$ and, following the same pattern as above, obtain:

8.7. **Lemma.** Let $U \subset \mathbf{R}^2 = \{(x, z) | x, z \in \mathbf{R}\}$ be open. Let $f : U \to \mathbf{R}$ be differentiable and $K \subset U$ compact. There are functions $g : U \to \mathbf{R}$, arbitrarily close to f on K such that

(I) $\partial g/\partial z(a) \neq 0$, or

(II) $\partial^2 g/\partial z^2(a) \neq 0$, or

(III) $\partial^2 g/\partial x \partial z(a) \neq 0$ and $\partial^3 g/\partial z^3(a) \neq 0$, for all $a \in U$.

Proof. Put

$$g(x, z) = f(x, z) + \lambda_1 z + \lambda_2 z^2 + \lambda_3 xz + \lambda_4 z^3.$$

We have to show that $\lambda = (\lambda_1, \ldots, \lambda_4) \in \mathbf{R}^4$ may be chosen so small that either (I) or (II) or (III) is satisfied everywhere. We have

$$\partial g/\partial z = (\partial f/\partial z) + \lambda_1 + 2\lambda_2 z + \lambda_3 x + 3\lambda_4 z^2$$

$$\partial^2 g/\partial z^2 = (\partial^2 f/\partial z^2) + 2\lambda_2 + 6\lambda_4 z$$

$$\partial^2 g/\partial x \partial z = (\partial^2 f/\partial x \partial z) + \lambda_3$$

$$\partial^3 g/\partial z^3 = (\partial^3 f/\partial z^3) + 6\lambda_4.$$

It follows that if $\partial g/\partial z(a) = \partial^2 g/\partial z^2(a) = 0$, then

$$\partial^2 g/\partial x \partial z(a) = 0 \Longleftrightarrow A(a).\lambda = b(a)$$

where

$$A(a) = \begin{bmatrix} 1 & 2z & x & 3z^2 \\ 0 & 2 & 0 & 6z \\ 0 & 0 & 1 & 0 \end{bmatrix} \quad \text{and} \quad b(a) = -\left(\frac{\partial f}{\partial z}, \frac{\partial^2 f}{\partial z^2}, \frac{\partial^2 f}{\partial x \partial z}\right).$$

There is a corresponding condition for $\partial^3 g/\partial z^3(a) = 0$, differing only in the form of the matrix. In this case it is

$$\bar{A}(a) = \begin{bmatrix} 1 & 2z & x & 3z^2 \\ 0 & 2 & 0 & 6z \\ 0 & 0 & 0 & 6 \end{bmatrix}$$

The matrices A and $\bar{A}$ have rank 3 everywhere, it suffices to show that those $\lambda = (\lambda_1, \ldots, \lambda_4)$, for which $A(a).\lambda = b(a)$ for some a, form a thin set. And that the same is true for $\bar{A}$.

Now A defines a map

$$U \times \mathbf{R}^4 \to U \times \mathbf{R}^3$$
$$(a, \lambda) \mapsto (a, A(a).\lambda);$$

because A has rank 3, this map is a submersion. It has rank 5 everywhere.

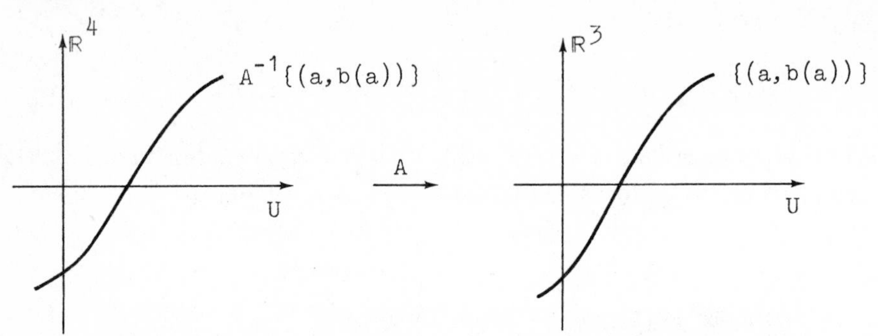

The points $\{(a, b(a)) | a \in U\}$ form a submanifold in $U \times \mathbf{R}^3$ of codimension 3. Its inverse image in $U \times \mathbf{R}^4$ is therefore also a submanifold of codimension (and hence dimension) 3. If one projects this manifold onto $\mathbf{R}^4$, one obtains a thin set consisting of those $\lambda \in \mathbf{R}^4$ for which $A(a). \lambda = b(a)$ for some $a \in U$. ✓

With the help of a locally finite covering, the standard procedure gives an open and dense subset $T \subset C^\infty(\mathbf{R}^2, \mathbf{R}^2)$ such that each map in T has, at each point, a local coordinate representation

$$(x, z) \mapsto (x, f(x, z))$$

where f satisfies one of the conditions (I), (II) or (III) of lemma 8. 7.

We shall outline how this works. By lemma 8. 5, there is an open and dense set of maps $F : \mathbf{R}^2 \to \mathbf{R}^2$ such that for each such F we may choose a countable family of compact sets $\{K_n | n \in \mathbf{N}\}$ and neighbourhoods U_n of K_n for each n with the following properties:

the interiors of the K_n cover $\mathbf{R}^2$,
the U_n have compact closure,
$\{U_n | n \in \mathbf{N}\}$ is locally finite,

and if $\mathbf{R}^2 = \{(x_1, x_2)\}$

for each n there is a pair (i, j) where $i, j \in \{1, 2\}$,

with $\partial F_i/\partial x_j(a) \neq 0$ for each $a \in K_n$.

It is possible to choose a fixed map for the whole of K_n $(x_1, x_2) \mapsto (x_k, F_i)$ which is locally a coordinate transformation at any point of K_n. This map transforms the original germ into the form $(x_1, x_2) \mapsto (x_1, f(x_1, x_2))$.

Observe that the last condition on K_n implies $\partial G_i/\partial x_j(a) \neq 0$ for all $a \in K_n$ when G is near enough to F. For such a G there is a coordinate transformation, corresponding to the one for F, which gives G the form $(x_1, x_2) \mapsto (x_1, g(x_1, x_2))$. If G is close to F then g is close to f.

Now alter the original map inductively on the U_n in such a way that after the n-th step the conditions of lemma 8.7 are satisfied on $\bigcup_{i=1}^{n} K_i$. Assume this has been achieved for $\bigcup_{i=1}^{n} K_i$ and that the map F has been altered so little that $\partial F_i/\partial x_j(a)$ is still non-zero for all $a \in K_m$ and for all m - where for each m the correct pair (i, j) is considered. Next change F on U_{n+1} so that the conditions of lemma 8.7 are satisfied on K_{n+1} but by such a small amount that (1) the map stays in a prescribed neighbourhood of the original map, (2) the conditions of lemma 8.7 are retained on $\bigcup_{i=1}^{n} K_n \cap U_{n+1}$ and (3) $\partial F_i/\partial x_j(a) \neq 0$ for all $a \in K_m$, all m and suitable (i, j).

Because the family $\{U_n | n \in \mathbf{N}\}$ is locally finite, there are locally only finitely many changes in F. The limit of the above procedure is therefore a differentiable map which can be locally transformed into the form $(x, y) \mapsto (x, f(x, y))$ where f satisfies the conditions of lemma 8.7.

The second step in the proof of Whitney's theorem will be to show that a map in T can be transformed locally into one of the three forms considered above: a regular point, a fold or a cusp.

Thus we consider a germ

$$\begin{aligned} \tilde{F} : (\mathbf{R}^2, 0) &\to (\mathbf{R}^2, 0) \\ (x, z) &\mapsto (x, \tilde{f}(x, z)) \end{aligned}$$

where

(I) $\partial f/\partial z(0) \neq 0$ or

(II) $\partial f/\partial z(0) = 0$ and $\partial^2 f/\partial z^2(0) \neq 0$, or

(III) $\partial f/\partial z(0) = \partial^2 f/\partial z^2(0) = 0$ and
$\partial^2 f/\partial x \partial z(0) \neq 0$, $\partial^3 f/\partial z^3(0) \neq 0$.

8.8. **Case (I).** $\tilde{F}$ is regular. Transforming the image space locally using F^{-1} makes F the identity.

8.9. **Case (II).** $f(0, z) = z^2 . q(z)$ with $q(0) \neq 0$. Hence the ideals $\langle x, f\rangle_{\mathcal{E}(2)}$, $\langle x, z^2\rangle_{\mathcal{E}(2)}$ are equal and, by the preparation theorem, the functions $1, z$ are generators of $\mathcal{E}(2)$ as an $\mathcal{E}(2)$-module via F^*.

In particular, the germ z^2 can be written

$$z^2 = \Phi(x, f(x, z)) + 2\psi(x, f(x, z)). z$$

for some germs $\Phi(x, y)$, $\psi(x, y)$.

The second order Taylor expansion of this identity is an identity between second order polynomials and the coefficients are identical

$$\Phi(0, 0) = \psi(0, 0) = 0; \quad \partial\Phi/\partial y(0) \neq 0;$$

$$\partial\psi(x, f(x, z))/\partial z(0) = \partial\psi/\partial y(0). \partial f/\partial z(0) = 0.$$

This means that h and k are coordinate transformations where

$$h(x, z) = (x, z - \psi(x, f(x, z)))$$
$$k(x, y) = (x, \Phi(x, y) + \psi^2(x, y)).$$

The following diagram is now commutative:

$$\begin{array}{ccc}
(x, z) & \overset{h}{\longmapsto} & (x, z - \psi(x, f(x, z))) \\
\Big\downarrow F & & \Big\downarrow (\mathrm{id}, \mathrm{id}^2) \\
(x, f(x, z)) & \overset{k}{\longmapsto} & (x, \Phi(x, f(x, z)) + \psi^2(x, f(x, z))),
\end{array}$$

since $z^2 + \psi^2 - 2\psi z = \Phi + 2\psi z + \psi^2 - 2\psi z = \Phi + \psi^2$. Hence F is equivalent to $(x, z) \mapsto (x, z^2)$.

8.10. **Case (III).** Here $\partial f/\partial z(0) = \partial^2 f/\partial z^2(0) = 0$ and neither $\partial^2 f/\partial x\partial z(0)$ nor $\partial^3 f/\partial z^3(0)$ vanish. As before, we deduce from the preparation theorem that there are germs $\tilde{\Phi}, \tilde{\psi}, \tilde{\theta}$ such that

$$z^3 = \tilde{\Phi}(x, \tilde{f}(x, z)) + \tilde{\psi}(x, \tilde{f}(x, z))z + 3\tilde{\theta}(x, \tilde{f}(x, z))z^2.$$

This identity induces an identity between jets at the origin. By comparing coefficients of 1, z, z^2, one finds $\tilde{\Phi}(0) = \tilde{\psi}(0) = \tilde{\theta}(0) = 0$, and $\theta(0, f(0, z))$ vanishes to at least third order in z (recall that this means that it lies in $\mathfrak{m}(1)^3$). Hence

$$(x, z) \mapsto (x, z - \tilde{\theta}(x, \tilde{f}(x, z)))$$

is a permissible coordinate transformation.

By this change of coordinates f transforms to a function f_T given by the following diagram:

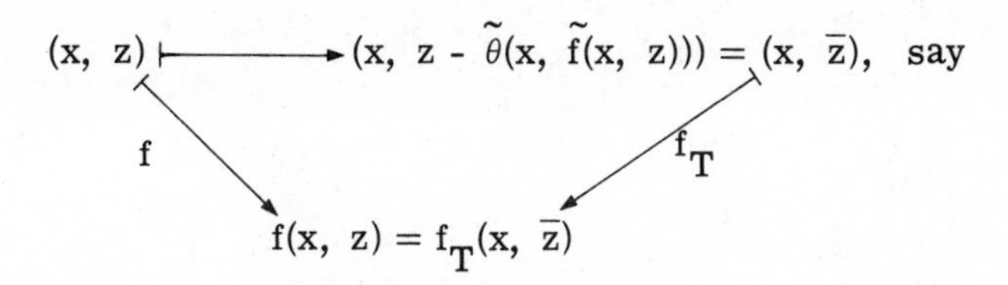

The function f_T also satisfies the condition for f in case (III): the function $f_T(0, \bar{z}) = f(0, z - \theta(0, f(0, z)))$ vanishes to exactly third order, and $\partial f_T/\partial x(0, \bar{z})$ vanishes exactly to first order, as can be seen by comparing coefficients of 2-jets:

$$\hat{f}(x, z) = f_1 x + f_2 x^2 + f_3 xz \mod \hat{\mathfrak{m}}(2)^3$$

$$\hat{\bar{z}}(x, z) = z + c_1 x + c_2 x^2 + c_3 xz \mod \hat{\mathfrak{m}}(2)^3, \text{ hence}$$

$$\hat{z}(x, \bar{z}) = \bar{z} - c_1 x - (c_2 - c_1 c_3)x^2 - c_3 x\bar{z} \mod \hat{\mathfrak{m}}(2)^3, \text{ hence}$$

$$\hat{f}_T(x, \bar{z}) = \hat{f}(x, \hat{z}(x, \bar{z}))$$

$$= f_1 x + (f_2 - f_3 c_1)x^2 + f_3 x\bar{z} \mod \hat{\mathfrak{m}}(2)^3$$

f_3 is non-zero by hypothesis.

Now consider that in the new coordinates

$$\bar{z}^3 = (z - \theta)^3 = z^3 - 3z^2\theta + 3z\theta^2 - \theta^3$$
$$= \Phi + \psi z + 3\theta z^2 - 3z^2\theta + 3z\theta^2 - \theta^3$$
$$= (\psi + 3\theta^2)(z - \theta) + (\Phi + 2\theta^3 + \psi.\theta)$$
$$= \psi_1 \bar{z} + \Phi_1, \text{ say.}$$

This shows that we could have assumed $\theta = 0$ from the beginning, that is:

$$z^3 = \Phi(x, f) + \psi(x, f)z;$$
$$\Phi(0, 0) = \psi(0, 0) = 0.$$

Next, make the coordinate changes in the domain $\{(x_1, x_2)\}$ and the range $\{(y_1, y_2)\}$ given by

$$\text{(A)} \quad \begin{bmatrix} x_1 \\ x_2 \end{bmatrix} \to \begin{bmatrix} \psi(x_1, f(x_1, x_2)) \\ x_2 \end{bmatrix}$$

$$\text{(B)} \quad \begin{bmatrix} y_1 \\ y_2 \end{bmatrix} \to \begin{bmatrix} \psi(y_1, y_2) \\ \Phi(y_1, y_2) \end{bmatrix} .$$

First we must check that these are valid as changes of coordinates. This is a computation with 3-jets:

$$\hat{f}(x_1, x_2) = f_1x_1 + f_2x_1^2 + f_3x_1x_2 + f_4x_2^3 + x_1.o(2) \mod \hat{m}^4,$$

with $f_3, f_4 \neq 0$ (recall the $o(\ )$ notation introduced in 4.7)

$$\hat{\Phi}(x_1, y) = b_1x_1 + b_2y \mod \hat{m}^2$$
$$\hat{\psi}(x_1, y) = c_1x_1 + c_2y \mod \hat{m}^2, \text{ and}$$
$$\text{(C)} \quad x_2^3 = \hat{\Phi}(x_1, \hat{f}(x_1, x_2)) + \hat{\psi}(x_1, \hat{f}(x_1, x_2)).x_2 .$$

For (A) we must show that $c_1 + c_2f_1 \neq 0$, and for (B) that $\begin{vmatrix} b_1 & b_2 \\ c_1 & c_2 \end{vmatrix} \neq 0.$

It follows from (C), which is an identity between power series, that $b_2 \neq 0$, since b_2f_4 is the coefficient of x_2^3 on the right. On the other hand the coefficient of x_1x_2, namely $(c_1 + c_2f_1) + b_2f_3$, is zero, and because $f_3 \neq 0$, $b_2 \neq 0$, we must have $c_1 + c_2 f_1 \neq 0$. The coefficient of x_1 is $b_1 + b_2 f_1$, this is zero so that $b_1 c_2 - b_2 c_1 = -b_2(f_1 c_2 + c_1) \neq 0$, as required.

To complete Whitney's theorem on mappings of the plane into the plane, we must check the commutativity of the following diagram

$$\begin{array}{ccc} (x_1, x_2) & \xmapsto{(A)} & (\psi(x_1, f(x_1, x_2)), x_2) = (\bar{x}, \bar{y}), \text{ say} \\ F \downarrow & & \downarrow \\ (x_1, f(x_1, x_2)) & \xmapsto{(B)} & (\psi(x_1, f(x_1, x_2)), \Phi(x_1, f(x_1, x_2))) \\ & & = (\bar{x}, \bar{y}^3 - \overline{xy}) \end{array}$$

which is equivalent to our equation:

$$x_2^3 - \psi(x_1, f(x_1, x_2))x_2 = \Phi(x_1, f(x_1, x_2)). \checkmark$$

9 · Boardman-Thom singularities

Literature: R. Thom and H. Levine: Singularities of differentiable mappings, Bonn, Math. Institut 1959, republished in: Liverpool Singularities - Symposium I, Lecture Notes 192, Springer (1971), 1-89.

J. M. Boardman: Singularities of differentiable maps, I. H. E. S. Publ. Math., 33 (1967), 21-57.

J. Mather: On Thom-Boardman singularities, Internat. Sympos. in Dynamical Systems (Salvador 1971), Academic Press, New York (1973).

J. Milnor: Differential topology, lecture notes, Princeton (1958).

Let $f : M \to N$ be a differentiable map. As one of the exercises has already shown, the set

$$\Sigma^i(f) = \{x \in M \mid \dim(M) - \mathrm{rk}_x f = i\} \subset M$$

may be very wild, in certain cases it could be any closed subset of M. On the other hand, for mappings of surfaces into surfaces the set $\Sigma^i(f)$ is a manifold for almost any f. For Whitney's maps only the sets $\Sigma^0(f)$ and $\Sigma^1(f)$ appear, the first is open and the second consists of differentiable curves.

If $\Sigma^i(f)$ is a manifold, then one may define the set $\Sigma^{i,j}(f) = \Sigma^j(f|\Sigma^i(f))$. In the case of Whitney's maps the only interesting set is $\Sigma^{1,1}(f)$ which is the set of cusp points of f. As long as manifolds are obtained the construction may be iterated by inductively defining the set $\Sigma^{i_1, i_2, \ldots, i_n}(f)$. The question, posed by Thom and subsequently answered affirmatively by Boardman, is: does there exist an open and dense subset $T \subset C^\infty(M, N)$, so that for all $f \in T$, all the subsets

$\Sigma^{i_1, i_2, \ldots, i_n}(f)$ can be defined in the way described and are submanifolds? In fact, Boardman defined a residual subset T (residual means a countable intersection of open, dense sets).

Mather has remarked that if M is compact then this T contains an open, dense set with the same properties - see the cited reference §6, proposition 2.

In order that the book should remain an introduction to the theory of differentiable maps and not get involved in all of Boardman's results, we shall only consider the singularity subsets $\Sigma^i(f)$. This case was dealt with by Thom.

Almost every theorem about 'almost every' map depends on arguments about 'general position' with respect to certain submanifolds. In differential topology, the idea of 'general position' is described by transversality. First we describe this concept.

Let $f : M^m \to N^n$ be differentiable and $U^{n-k} \subset N^n$ a submanifold of codimension k:

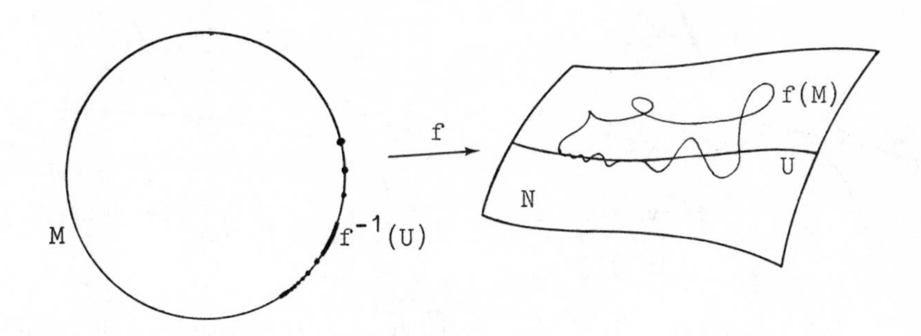

It is not necessarily true that $f^{-1}(U) \subset M$ is a submanifold. We already know that if $U = \{0\} \subset \mathbf{R}^1$, then $f^{-1}(U)$ can be an arbitrary closed subset. It is easy to see this result is valid more generally.

9.1. Definition. The map $f : M^m \to N^n$ is called transversal to the submanifold $U^{n-k} \subset N^n$, if the following condition is satisfied for every $x \in M$ such that $f(x) \in U$:

Let $(y_1, \ldots, y_n)$ be local coordinates at f(x), such that near f(x) the submanifold U is defined by $y_1 = y_2 = \ldots = y_k = 0$ (see the definition of submanifold). Then the condition is that the matrix

$$(\partial f_i/\partial x_j) \quad 1 \le i \le k, \quad 1 \le j \le m$$

has rank k at the point x, where $(x_1, \ldots, x_m)$ is a coordinate system at $x \in M$.

Note that this condition is only a restriction on those $x \in M$ for which $f(x) \in U$. Hence if $m < k$, transversality means that $f(M) \cap U = \emptyset$.

Using the tangent spaces and the tangent map, the condition may be described as follows:

If $x \in M$ and $f(x) \in U$, then the map

$$T_xM \xrightarrow{T_xf} T_{f(x)}N \longrightarrow T_{f(x)}N/T_{f(x)}U$$

is surjective. Alternatively

$$T_xf(T_xM) + T_{f(x)}U = T_{f(x)}N$$

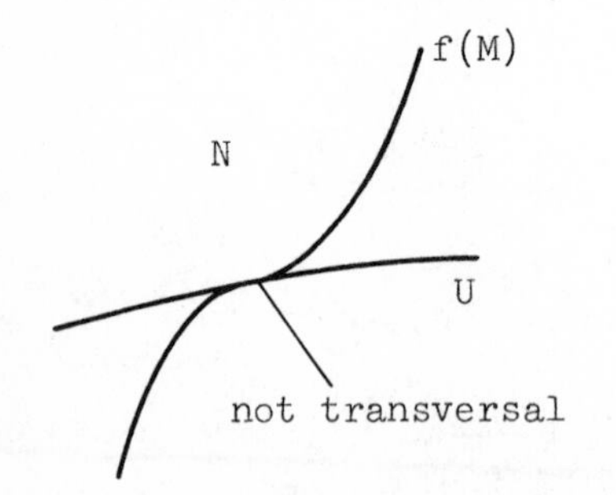

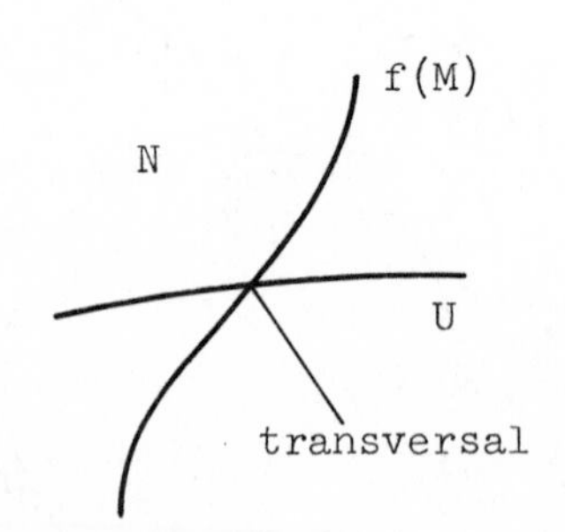

9.2. **Remark.** If $f : M^m \to N^n$ is transversal to the submanifold $U^{n-k} \subset N^n$, then $f^{-1}(U) \subset M$ is a submanifold with codimension k.

Proof. Choose coordinate systems $(x_1, \ldots, x_m)$ and $(y_1, \ldots, y_n)$ as in the definition of transversality. Near $x \in f^{-1}(U)$ the set $f^{-1}(U)$ is locally the inverse image of zero under the composite map

$$(x_1, \ldots, x_m) \overset{f}{\mapsto} (y_1, \ldots, y_n) \overset{\text{proj}}{\mapsto} (y_1, \ldots, y_k).$$

This map has rank k, by definition of transversality. ✓

If $U \subset N$ is closed, then the set of maps which are transversal to U is an open subset of $C^{\infty}(M, N)$. This is because transversality is expressed, locally, by the non-vanishing of a continuous map $\phi_f : M \to \mathbf{R}$, namely the sum of the squares of the distance $(f(x), U)$ and the determinants $(\partial f_i / \partial x_j)$, $i \leq k$. If f changes only slightly, then so does ϕ_f.

On the other hand Thom's transversality lemma (see Milnor's lectures on differentiable topology) states that the set of maps transversal to U is dense in $C^{\infty}(M, N)$.

The local part of this theorem follows directly from Sard's theorem. If $f : M \to \mathbf{R}^n$ is differentiable, let $\lambda = (\lambda_1, \ldots, \lambda_k)$ be a regular value of the composite

$$M \xrightarrow{f} \mathbf{R}^n \xrightarrow{p} \mathbf{R}^k$$

where $p(y_1, \ldots, y_n) = (y_1, \ldots, y_k)$. Then the map $M \to \mathbf{R}^n$ given by

$$x \to f(x) - (\lambda_1, \ldots, \lambda_k, 0, \ldots, 0)$$

is near f for small λ and transversal to

$$\mathbf{R}^{n-k} = \{(y_1, \ldots, y_n) \in \mathbf{R}^n | y_1 = \ldots = y_k = 0\}.$$

The transition from the local to the global will not be repeated, it depends on the argument with locally finite coverings which was indicated at the ends of the proofs of lemmas 8.5 and 8.7.

After these preliminaries about general position, we return to singularity sets. First, here is the solution to an exercise which was set earlier.

9.3. **Lemma.** Let $LA(n, m) = Hom(\mathbf{R}^n, \mathbf{R}^m)$ be the vector space of real $(m \times n)$-matrices, and $LA(n, m; r) \subset LA(n, m)$ the subspace of matrices with rank r. Then $LA(n, m; r)$ is a differentiable submanifold of $LA(n, m)$ with codimension $(n-r).(m-r)$, for $r \leq \min(m, n)$.

Proof. Let $U \subset LA(n, m)$ be the open subset of matrices of the form

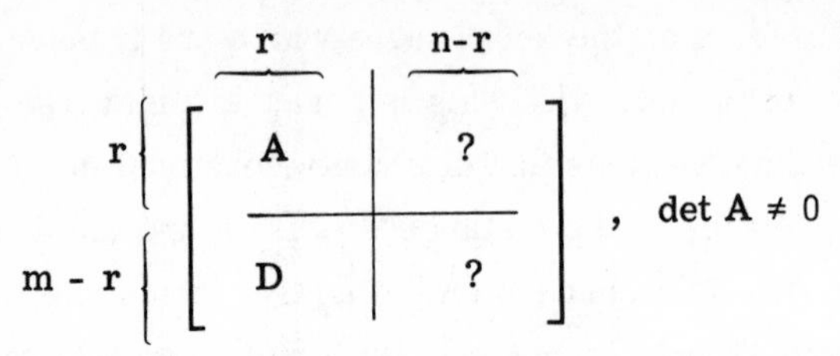

A point in LA(n, m; r) can be assumed to lie in U (otherwise, reorder the rows and columns). In this case $U \cap LA(n, m; r)$ consists of those matrices where the last $n - r$ columns depend on the first r. Now take matrices A, B, C, D as follows

$A : (r \times r)$-matrix, $\det A \neq 0$
$B : (r \times (n - r))$-matrix
$C : ((m - r) \times (n - r))$-matrix
$D : ((m - r) \times r)$-matrix

and consider the map

$$(A, B, C, D) \mapsto \left[\begin{array}{c|ccc} A & & 0 & \\ \hline & 1 & & 0 \\ D & & \ddots & \\ & 0 & & 1 \end{array}\right] \cdot \left[\begin{array}{ccc|c} 1 & & 0 & B \\ & \ddots & & \\ 0 & & 1 & \\ \hline & 0 & & C \end{array}\right] = \left[\begin{array}{c|c} A & AB \\ \hline D & DB+C \end{array}\right]$$

(In the two factors, the left block columns have width r; the right block columns have widths $m-r$ and $n-r$ respectively.)

The image lies in U and the first matrix in the product is invertible since $\det A \neq 0$. Obviously this map is invertible, since A^{-1} exists. The image of (A, B, C, D) lies in LA(n, m; r) if and only if $C = 0$. This part of U is a submanifold of codimension $(m-r).(n-r)$ in the manifold of all matrices (A, B, C, D) as above. √

9.4. Lemma. Let $f : \mathbf{R}^n \to \mathbf{R}^m$ be differentiable and $U \subset LA(n, m)$ a differentiable submanifold, then for almost any linear map $A : \mathbf{R}^n \to \mathbf{R}^m$ (i.e. for almost any matrix) the map

$$\mathbf{R}^n \to LA(n, m); \; x \mapsto D(f + A)(x) = Df(x) + A$$

is transversal to U.

Proof. Consider the map

$$\phi : \mathbf{R}^n \times U \to LA(n, m); \quad (x, u) \mapsto u - Df(x).$$

Let $A \subset LA(n, m)$ be a regular value of ϕ, then we claim that the map $\mathbf{R}^n \to LA(n, m)$; $x \mapsto Df(x) + A$ is transversal to U. If u is a point of U and $u = Df(x) + A$, then $u - Df(x) = A$ is a regular value of ϕ (and a value of ϕ) so that near (x, u), ϕ is a submersion (it has rank n. m, and the tangent map $T\phi$ is an epimorphism). This means that the image of the tangent map to Df and the tangent space to U at u generate the tangent space to $LA(n, m)$ at A. Notice that the differential of the map $Df : \mathbf{R}^n \to LA(n, m)$ is the same as the differential of the map $Df + A$, since A is a constant in the vector space $LA(n, m)$. √

These lemmas give us the following

9.5. **Theorem (Thom).** <u>For a differentiable map</u> $f : \mathbf{R}^n \to \mathbf{R}^m$, <u>let</u>

$$\Sigma^r(f) = \{x \in \mathbf{R}^n | Rk_x f = n - r\}.$$

<u>For almost every linear map</u> $A : \mathbf{R}^n \to \mathbf{R}^m$ <u>the set</u> $\Sigma^r(f + A)$ <u>is a differentiable submanifold of</u> $\mathbf{R}^n$ <u>of codimension</u> $(m - n + r). r$, <u>for every</u> r.

Proof. $\Sigma^r(f) = Df^{-1}(LA(n, m; n-r))$, and $LA(n, m; n-r)$ is a submanifold of $LA(n, m)$ with codimension $(m-(n-r)). (n-(n-r))=(m-n+r). r$ (Lemma 9. 3). By lemma 9. 4, f can be deformed, by the addition of almost any linear map A, in such a way that $D(f + A)$ is transversal to every $LA(n, m; n-r)$. By our earlier remark (9. 2), $\Sigma^r(f + A)$ is a submanifold with the same codimension as $LA(n, m; n-r)$. √

Using our standard argument with locally finite coverings we may deduce the more general result that for an open and dense set of maps $f \in C^\infty(M^m, N^n)$, all the sets $\Sigma^r(f) = \{x \in M | Rk_x f = n - r\}$ are submanifolds. An additional complication is that the manifolds $LA(n, m; r)$ are not closed. However the aim, here, is only to create an interest in these questions, and to outline the available literature.

Boardman's handling of the general case $\Sigma^{i_1,\ldots,i_r}(f)$ depends on defining certain algebraic, regular manifolds $\Sigma(i_1, \ldots, i_r)$ in the jet space $\hat{\mathscr{E}}(n, m)$ (instead of the space LA(n, m) of 1-jets). Then he deforms a given map $f : \mathbf{R}^n \to \mathbf{R}^m$, slightly, so that the map

$$jf : \mathbf{R}^n \to \hat{\mathscr{E}}(n, m)$$
$$x \mapsto (\text{the jet of the germ } f : (\mathbf{R}^n, x) \to \mathbf{R}^m \text{ at } x)$$

is transversal to all the submanifolds $\Sigma(i_1, \ldots, i_r)$. For such maps, he shows that

$$\Sigma^{i_1,\ldots,i_r}(f) = jf^{-1}\Sigma(i_1, \ldots, i_r)$$

and that each is a submanifold.

10 · The quadratic differential

Literature: V. I. Arnol'd: Singularities of smooth mappings, Russian Math. Surveys, 23 (1969), 1-43 (translated from Uspehi Mat. Nauk, 23 (1968), 3-44).
S. Lang: Analysis I, Addison-Wesley (1968).
I. R. Porteous: Simple singularities of maps, lectures, Columbia 1962, Liverpool Singularities - Symposium I, Lecture notes 192, Springer (1971), 286-307.
R. Thom: La stabilité topologique des applications polynomiales, l'Enseignement Math. 8 (1962), 24-33.

In the case of maps between surfaces, the stable differentiable maps form an open and dense subset in the set of all maps. There is no corresponding result for manifolds of sufficiently high dimension. The stable maps - in the sense of differentiable equivalence - are not dense.

To prove this we shall define invariants which help us to decide when germs are non-equivalent.

Roughly speaking, these invariants (introduced by Porteous) are given by that quadratic part of the Taylor expansion where the linear part vanishes.

This quadratic part in the germ's Taylor expansion defines certain linear families of quadratic forms. By algebraic means, these forms can be divided into equivalence classes. Hence we obtain invariants for the linear families and, in turn, for germs.

We begin with details of the quadratic differential.

Let $f : (\mathbf{R}^n, 0) \to (\mathbf{R}^m, 0)$ be a differentiable germ. The differential of f is a linear map $Df(0) : \mathbf{R}^n \to \mathbf{R}^m$, where $\mathbf{R}^n$ and $\mathbf{R}^m$ are canonically identified with their tangent spaces at the origin.

10.1. The quadratic differential is a quadratic map

$$d^2f_0 : \mathrm{Ker}(Df(0)) \to \mathrm{Coker}(Df(0))$$

defined as follows

$$d^2f_0(v) = \lim_{t\to 0} \frac{f(t.v)}{t^2} \text{ modulo } Df(0)(\mathbf{R}^n)$$

for $v \in \mathrm{Ker}(Df(0))$.

We have to show that this limit exists and that the map d^2f_0 is well defined as a quadratic map:

$$\mathrm{Ker}(Df(0)) \to \mathrm{Coker}(Df(0))$$

where these vector spaces are regarded as subspaces of the tangent spaces at the origin to $\mathbf{R}^n$ and $\mathbf{R}^m$ respectively. For this we have to show that d^2f_0 is independent of differentiable coordinate changes.

Consider the Taylor expansion

$$\text{(1)} \qquad f(tv) = tDf(0).\, v + \tfrac{1}{2}t^2 \sum_{j,k} \frac{\partial^2 f}{\partial v_j \partial v_k} \,.\, v_j v_k + o(3).$$

Let $v \in \mathrm{Ker}(Df(0))$. The first term vanishes and so the required limit exists and is equal to

$$\text{(2)} \qquad \tfrac{1}{2} \sum_{j,k} \frac{\partial^2 f}{\partial v_j \partial v_k} \,.$$

If $\phi : (\mathbf{R}^m, 0) \to (\mathbf{R}^m, 0)$ is a diffeomorphism, then

$$d^2(\phi \circ f)_0(v) = D\phi(0) \,.\, \tfrac{1}{2} \sum_{j,k} \frac{\partial^2 f}{\partial v_j \partial v_k} \,.\, v_j v_k$$

using the Taylor expansion (1). Hence d^2f_0 is transformed by the Jacobian matrix in exactly the same way as the tangent space at the origin to $\mathbf{R}^m$. Hence the map is independent of coordinate changes on $\mathbf{R}^m$.

Next if $\psi : (\mathbf{R}^n, 0) \to (\mathbf{R}^n, 0)$ is a coordinate transformation at the origin with $D\psi(0) = \mathrm{id}$, then $\psi(tv) = tv + t^2 w(t)$ and

$$f(\psi(tv)) = t^2 Df(0).\, w(t) + \tfrac{1}{2}t^2 \sum_{j,k} \frac{\partial^2 f}{\partial v_j \partial v_k} v_j v_k + o(3),$$

hence

$$\lim_{t\to 0} \frac{f.\psi(tv)}{t^2} = \lim_{t\to 0} \frac{f(tv)}{t^2} + Df(0).\,w(0).$$

The last term belongs to $Df(0)(\mathbf{R}^n)$.

It has now been shown that the map d^2f_0 is uniquely defined by equation (2) once a basis has been given for the subspace $\operatorname{Ker} Df(0)$ of the tangent space to $\mathbf{R}^n$ at the origin.

10.2. Example. The map $f : \mathbf{R}^4 \to \mathbf{R}^4$

$$(x_1, x_2, x_3, x_4) \mapsto (x_1, x_2, x_3^2 - x_4^2 + x_1x_3 + x_2x_4, x_3x_4)$$

has the quadratic differential

$$(x_3, x_4) \mapsto (x_3^2 - x_4^2, x_3x_4).$$

In particular it is clear that the quadratic differential can have any given form by making a suitable choice of m, n, $\dim(\operatorname{Ker} Df(0))$ and a polynomial map f.

A pencil of quadratic forms can be assigned to the quadratic differential in an invariant way. Define:

F = the vector space of quadratic forms on $\operatorname{Ker} Df(0)$

C = the dual space to $\operatorname{Coker} Df(0)$

$L_f : C \to F$, $L_f(\alpha) = \alpha \circ d^2f_0$.

10.3. Definition. The map $L_f : C \to F$ is called the <u>pencil of quadratic forms</u> for d^2f_0.

10.4. Example. Assume $\operatorname{Ker} Df(0)$ and $\operatorname{Coker} Df(0)$ both have dimension 2. The dimension of F is then 3 and the basis is $\{x^2, 2xy, y^2\}$ where x, y are coordinates for $\operatorname{Ker} Df(0)$. The form $ax^2 + 2bxy + cy^2$ corresponds to a point $(a, b, c) \in F$ and has matrix

$$A = \begin{pmatrix} a & b \\ b & c \end{pmatrix}$$

as a map from $\operatorname{Ker} Df(0)$ to its dual space. The matrix $A = 0$ corresponds to the zero quadratic form. If $A \neq 0$, the matrices with $ac - b^2 = 0$ correspond to forms of rank 1. The normal form for these is $\pm x^2$. When $\det A \neq 0$ the normal forms are $x^2 \pm y^2$ and $-(x^2 + y^2)$.

The set $\{\det A = 0\}$ is a cone in F:

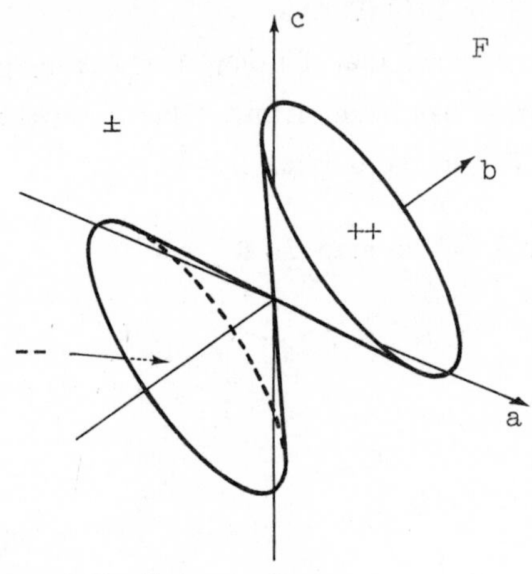

The vertex of the cone is the zero form.

The points of the cone are the parabolic forms.

The regions ++, -- are elliptic forms.

The region ± is hyperbolic.

The isomorphism type of a quadratic form is determined by its position in relation to this cone.

If $L_f : C \to F$ is a pencil of quadratic forms, then $L_f(C) \subset F$ can lie in seven different ways:

1. a plane outside the cone,
2. a plane cutting the cone,
3. a plane tangent to the cone,
4. a line inside the cone,
5. a line outside the cone,
6. a line tangent to the cone,
7. the vertex.

Suppose two differentiable germs have corresponding pencils of quadratic forms such that the positions for $L_f(C)$ are different according to this classification. The germs cannot be equivalent because transformations of coordinates in domain and range give rise to coordinate changes for the pencil L_f and these cannot change the position of $L_f(C)$

in relation to the cone $\{\det A = 0\}$. It is in this way that the pencil of quadratic forms gives an algebraic invariant for the differentiable equivalence class of a germ.

10.5. More generally let $F(k)$ = the vector space of quadratic forms on $\mathbf{R}^k$, $H(c, k) = LA(\mathbf{R}^c, F(k))$ the vector space of c-dimensional pencils of quadratic forms on $\mathbf{R}^k$.

We must examine the operation of the general linear groups (linear coordinate changes on $\mathbf{R}^c$ and $\mathbf{R}^k$) on such pencils of forms in order to find the invariants for the elements of $H(c, k)$. The group $GL(c, \mathbf{R}) \times GL(k, \mathbf{R})$ of coordinate transformations on both $\mathbf{R}^c$ and $\mathbf{R}^k$ operates on $H(c, k)$ by the formula

$$g_c \times g_k(L_c(v_c)(v_k)) = L(g_c^{-1} v_c)(g_k^{-1} v_k).$$

For example we have worked out that $H(2, 2)$ has dimension 6 (since $\dim F(2) = 3$), and falls into 7 <u>orbits</u> under the action of $GL(2, \mathbf{R}) \times GL(2, \mathbf{R})$). An orbit is an equivalence class of the operation of the group. In addition it is clear that $\dim(GL(2, \mathbf{R}) \times GL(2, \mathbf{R})) = 8$.

In definition 10.3, we assigned a pencil of quadratic forms $L \in LA(C, F)$ to each germ $f : (M^m, x) \to (N^n, y)$ using $d^2 f_0$. By choosing coordinates on C and F we obtain a unique element in $H(c, k)$. When new coordinates are chosen for C and F, this element in $H(c, k)$ is transformed using an element of $GL(c, \mathbf{R}) \times GL(k, \mathbf{R})$. Hence the orbit of the pencil L in $H(c, k)$ is a well defined invariant of the germ f up to differentiable equivalence. This leads to the following result:

10.6. **Theorem (R. Thom).** <u>The set of differentiably stable maps</u> <u>$M^{n^2} \to N^{n^2}$ is not dense in the set of all differentiable maps, when</u> $n \geq 3$.

Proof. The dimension of $GL(n, \mathbf{R}) \times GL(n, \mathbf{R})$ is $2n^2$. The dimension of $F(n)$ is the same as that of the symmetric $(n \times n)$-matrices, namely $\frac{1}{2}n(n + 1)$. Hence $\dim(H(n, n)) = \frac{1}{2}n^2(n + 1)$. For $n \geq 3$, we have $\frac{1}{2}(n + 1) \geq 2$ so that

$$\dim(GL(n, \mathbf{R}) \times GL(n, \mathbf{R})) \leq \dim(H(n, n)).$$

Equality only holds when $n = 3$ and for this case we observe that there is a one-dimensional subgroup of pairs (α, α^{-1}) of scalars which operates on the pencils as the identity. We deduce that, for $n \geq 3$, the orbits of the operation of $GL(n, \mathbf{R}) \times GL(n, \mathbf{R})$ have dimension smaller than that of $H(n, n)$, that is, a codimension ≥ 1.

Now consider a map $f : M^{n^2} \to N^{n^2}$, which has singularity type Σ^n at the point $0 \in M$ (this means $0 \in \Sigma^n(f)$). Suppose also that the differential of f defines a map transversal to $LA(n^2, n^2; n^2 - n)$ on a neighbourhood of 0 when local euclidean coordinates are introduced.

These two conditions that we have imposed on f are conditions on its first and second derivative. Thus the requirements can be satisfied using a polynomial of second order. As in the proof of Thom's theorem in chapter 9 we know that locally, in a coordinate system, $\Sigma^n(f) = Df^{-1}(LA(n^2, n^2; n^2 - n))$. From lemma 9.3 $\operatorname{codim} LA(n^2, n^2; n^2 - n) = n^2$ and so the transversality condition on f implies that $\dim \Sigma^n(f) = 0$.

Pick coordinates $(y_1, \ldots, y_{n^4})$ around $Df(0)$ on the manifold $LA(n^2, n^2)$ in such a way that the submanifold $LA(n^2, n^2; n^2 - n)$ is defined by $y_1 = \ldots = y_{n^2} = 0$. By transversality the derivative of the following composite at 0 is an isomorphism:

$$U \xrightarrow{Df} \{(y_1, \ldots, y_{n^4})\} \xrightarrow{\text{proj}} \{(y_1, \ldots, y_{n^2})\},$$

where U is a coordinate neighbourhood of 0. Hence if U is small enough, this composite will be a diffeomorphism and $0 = \Sigma^n(f) \cap U = (\mathrm{pr} \circ Df)^{-1}(0) \cap U$.

If f is slightly deformed, the composite will remain a diffeomorphism on a smaller neighbourhood $U' \subset U$ around 0. (Use Lang XIII, §3, lemma, p. 350, to see that U' does not depend on the particular approximation of f, provided the approximation is good enough.) If h is an approximation to f which differs only by a homogeneous polynomial of degree two then $0 \in \Sigma^n(h)$. In any case $\Sigma^n(h)$ will be a single point in U' which lies near 0. Hence if the stable maps were dense we should be able to find a stable map f with the properties we have just described. Now 0 is an isolated point of $\Sigma^n(f)$ and we may find an h arbitrarily close to f using a change in the second order term which makes d^2h_0 and d^2f_0 lie in different orbits of $H(n, n)$ under the action of $GL(n, \mathbf{R}) \times GL(n, \mathbf{R})$. This is because these orbits are all thin in $H(n, n)$. Therefore $f|U$ is not stable.

To complete the argument globally, first choose a map f such that its differential is everywhere transversal to $LA(n^2, n^2; n^2 - n)$ and such that locally, around $0 \in M$, it looks like the map $f|U$ above. Choose the approximation h so that h and f differ only on U and so that the orbit of d^2h_0 is distinct from all those orbits which contain the quadratic differential of f at any one of the points of $\Sigma^n(f)$. Since $\Sigma^n(f)$ is a manifold of dimension zero there are countably many such orbits. √

At first sight, this result appears to destroy any hopes which had been placed on the concept of stability. However Thom has discovered (or demanded) that the topologically stable maps always form a dense subset of $C^\infty(M, N)$, when M is compact. Mather has either given or announced proofs for this result and corresponding local ones. It is hoped that these results will appear in a work on 'topological stability' - perhaps in Springer's series 'Ergebnisse der Mathematik'.

The theory of topological stability is based on the theory of stratified sets (the decomposition of sets into differentiable manifolds). We shall not go into the subject here and unfortunately there is no literature suitable for a students' text. However several works have appeared which indicate that this is an area which will interest analysts and topologists.

11 · Finitely determined germs

Literature: J. Mather: Stability of C^∞-mappings III: finitely determined map-germs, I. H. E. S. Publ. Math., 35 (1968), 127-56.

J. C. Toujeron: Idéaux de fonctions différentiables I, Ann. Inst. Fourier, 18 (1968), 177-240.

R. C. Gunning and H. Rossi: Analytic functions of several complex variables, Prentice-Hall (1965).

J. Milnor: Singular points of complex hypersurfaces, Ann. of Math. Studies, 61 (1968).

As before, let $\mathcal{E}(n, m)$ be the vector space of germs $(\mathbf{R}^n, 0) \to \mathbf{R}^m$, and let $\mathcal{B}(n) \subset \mathcal{E}(n, n)$ be the subset of invertible germs (with respect to composition) with $f(0) = 0$. The pair $(\mathcal{B}(n), \circ)$ is then a group and $\tilde{f} \in \mathcal{E}(n, n)$ with $f(0) = 0$ is an element of $\mathcal{B}(n)$ if and only if $Df(0)$ is non-singular. The group $\mathcal{B}(n)$ acts on $\mathcal{E}(n, m)$ by composition: if $\tilde{f} \in \mathcal{E}(n, m)$ and $\tilde{h} \in \mathcal{B}(n)$, then $\tilde{f} \circ \tilde{h} \in \mathcal{E}(n, m)$.

11. 1. **Definition.** The germs $\tilde{f}_0, \tilde{f}_1 \in \mathcal{E}(n, m)$ are called right-equivalent if there is an $\tilde{h} \in \mathcal{B}(n)$ such that $\tilde{f}_0 \circ \tilde{h} = \tilde{f}_1$.

Let $j^k : \mathcal{E}(n) \to \mathcal{E}(n)/m(n)^{k+1} = \mathbf{R}[x_1, \ldots, x_n]/\langle x_1, \ldots, x_n\rangle^{k+1}$ be the jet map and for convenience, denote $\mathcal{E}(n)/m(n)^{k+1}$ by $\hat{\mathcal{E}}_k(n)$. To each germ $\tilde{f}$, the map j^k assigns the corresponding k-jet, the k-th Taylor polynomial of $\tilde{f}$ at the origin.

11. 2. **Definition.** A germ $\tilde{f} \in \mathcal{E}(n)$ is called k-determined (sufficient) if every germ $\tilde{g} \in \mathcal{E}(n)$, with the same k-jet as $\tilde{f}$, is right-equivalent to $\tilde{f}$.

This may be stated: if $j^k\tilde{f} = j^k\tilde{g}$, then there exists $\tilde{h} \in \mathcal{B}(n)$ such that $\tilde{f} = \tilde{g} \circ \tilde{h}$.

One may make the corresponding definition for germs in $\mathcal{E}(n, m)$, but this does not lead to any further, reasonable results. This is explained below, after we have become more familiar with the simplest kind of determinacy.

To say that $\tilde{f}$ is k-determined is much more a statement about the polynomial $j^k f$ than one about $\tilde{f}$. Each germ which has this polynomial as its k-jet looks like the polynomial when suitable coordinates are chosen. The k-jet determines the corresponding germ.

The question which is considered in this chapter is: Which jets in $\hat{\mathcal{E}}_k(n)$ determine the germs corresponding to them?

For example, note that

1. No 0-jet determines its germ.
2. $\hat{\mathcal{E}}_1(n) = \mathbf{R} \times \mathbf{R}^n$, since $j^1(\tilde{f}) = f_0 + \sum_i f_1^i x_i$, and $\tilde{f}$ is 1-determined if and only if some $f_1^i \neq 0$ (equivalently $Df(0) \neq 0$). This is because f may then be transformed into $(x_1, \ldots, x_n) \to x_1$.
3. The Morse-lemma states that if $f \in \mathcal{E}(n)$ and $Df(0) = 0$, then f is 2-determined if and only if

$$\det(\partial^2 f/\partial x_i \partial x_j(0)) \neq 0.$$

These examples show the non-determined jets become more scarce as the degree of the polynomial increases. The non-determined 1-jets form the line $\mathbf{R} \times \{0\}$ in $\mathbf{R}^{n+1}$. Over each point in the space of 1-jets there is the space of symmetric matrices which, going to 2-jets, represents all the possible coefficients of monomials of degree two. Above a point on the line of non-determined 1-jets the non-determined 2-jets also form a thin set given by $\{\det(a_{ij}) = 0\}$ in the space of symmetric matrices $\{(a_{ij})\}$.

In this chapter the intention is to prove the following result. It is one of many proved by Mather in this area and most of these are significantly stronger than this one.

11.3. Theorem. Let $\tilde{f} \in \mathcal{E}(n)$ and let

$$\mathfrak{m}(n)^k \subset \mathfrak{m}(n) \,.\, \left\langle \frac{\partial f}{\partial x_1}, \ldots, \frac{\partial f}{\partial x_n} \right\rangle_{\mathcal{E}(n)} + \mathfrak{m}(n)^{k+1},$$

then $\tilde{f}$ is k-determined.

First a few remarks:

The hypothesis may also be written

$$\mathfrak{m}(n)^k \subset \mathfrak{m}(n) \,.\, \langle \partial f/\partial x_i \rangle \text{ modulo } \mathfrak{m}(n)^{k+1}$$

($\langle \partial f/\partial x_i \rangle$ abbreviates $\langle \partial f/\partial x_1, \ldots, \partial f/\partial x_n \rangle_{\mathcal{E}(n)}$) and this is a condition on the k-jet of f - just as it should be.

If we write

$$\mathfrak{m}(n)^k \subset \mathfrak{m}(n)\langle \partial f/\partial x_i \rangle + \mathfrak{m}(n) \,.\, \mathfrak{m}(n)^k$$

then Nakayama's lemma implies

$$\mathfrak{m}(n)^k \subset \mathfrak{m}(n)\langle \partial f/\partial x_i \rangle_{\mathcal{E}(n)}.$$

This condition is therefore equivalent to the hypothesis in the theorem. However the formulation in the theorem has the advantage that only finite dimensional vector spaces are involved. This follows from the first remark, and for each given germ these vector spaces may be determined explicitly.

The last formulation of the condition shows that

$$\mathcal{E}(n)/(\mathfrak{m}(n) \,.\, \langle \partial f/\partial x_i \rangle)$$

is finite dimensional and generated by monomials of degree less than k. Conversely, let $\mathcal{E}(n)/(\mathfrak{m}(n).\langle \partial f/\partial x_i \rangle)$ have dimension k. Write $A = \mathcal{E}(n)/\mathfrak{m}(n)\langle \partial f/\partial x_i \rangle$. Using Nakayama's lemma we must have

$$0 = \mathfrak{m}(n)^l A \underset{\neq}{\subset} \mathfrak{m}(n)^{l-1} A \underset{\neq}{\subset} \ldots \underset{\neq}{\subset} \mathfrak{m}(n) A \underset{\neq}{\subset} A$$

where $l \leq k$ since $\dim A = k$. Hence

$$\mathfrak{m}(n)^k \subset \mathfrak{m}(n)^l \subset \mathfrak{m}(n)\langle \partial f/\partial x_i \rangle \,.$$

Further $\mathcal{E}(n) = \mathfrak{m}(n) \oplus \mathbf{R}$ where f corresponds to $(f - f(0)) \oplus f(0)$. Hence

$$\mathfrak{m}(n)\langle \partial f/\partial x_i \rangle + \langle \partial f/\partial x_i \rangle_{\mathbf{R}} = \langle \partial f/\partial x_i \rangle_{\mathcal{E}(n)}$$

and the condition that $\mathcal{E}(n)/\mathfrak{m}(n)\langle \partial f/\partial x_i \rangle$ has finite dimension is equivalent to $\mathcal{E}(n)/\langle \partial f/\partial x_i \rangle_{\mathcal{E}(n)}$ having finite dimension. Hence

11.4. If the germ $D\tilde{f} : (\mathbf{R}^n, 0) \to \mathbf{R}^n$ is finite (definition 6.8), then $\tilde{f}$ is finitely determined (k-determined for some k).

Using the argument with Nakayama's lemma given before, and this time setting $A = \mathcal{E}(n)/\langle \partial f/\partial x_i \rangle$, we find that if $\dim(\mathcal{E}(n)/(Df)^* \mathfrak{m}(n)\mathcal{E}(n)) = k$ then f is $(k+1)$-determined.

Next comes the proof of the theorem.

Let $\tilde{f}, \tilde{g} \in \mathcal{E}(n)$ be two germs with the same k-jet. We must show that there is a germ $\tilde{h} \in \mathcal{B}(n)$ with $\tilde{f} \circ \tilde{h} = \tilde{g}$. To do this, $\tilde{f}$ and $\tilde{g}$ are connected by the one-parameter family of germs $\tilde{F}$:

$$F(x, t) = (1 - t)f(x) + tg(x), \quad t \in \mathbf{R}, \ x \in \mathbf{R}^n.$$

Let $\tilde{F}_t \in \mathcal{E}(n)$ be given by $F_t(x) = F(x, t)$. We wish to show that for each $t_0 \in \mathbf{R}$, there is an $\varepsilon > 0$ such that $\tilde{F}_t$ is right-equivalent to $\tilde{F}_{t_0}$ whenever $|t - t_0| < \varepsilon$. This proves the theorem because $\mathbf{R}$ is connected. To prove this result about $\tilde{F}$ we look for a germ

$$\tilde{H} : (\mathbf{R}^n \times \mathbf{R}, (0, t_0)) \to \mathbf{R}^n,$$

which, denoting $\tilde{H}(x, t)$ by $\tilde{H}_t(x)$, has the properties

(I) $\quad \tilde{H}_{t_0} = \mathrm{id} \in \mathcal{B}(n)$

(II) $\quad \tilde{H}_t(0) = 0 \in \mathbf{R}^n$

(III) $\quad \tilde{F}_t \circ \tilde{H}_t = \tilde{F}_{t_0}$, that is, $F(H(x, t), t) = F(x, t_0)$.

For t near t_0 the germ $\tilde{H}_t$ is automatically invertible because $\tilde{H}_{t_0}$ is invertible and $\det(DH_t(0))$ depends continuously on t. Condition (III) is automatically fulfilled at $t = t_0$ and therefore it is sufficient to replace (III) by the differential equation which states that $F_t \circ H_t$ is independent of t, that is

(III') $$\sum_i \frac{\partial F}{\partial x_i}(H(x, t), t) \cdot \frac{\partial H_i}{\partial t}(x, t) + \frac{\partial F}{\partial t}(H(x, t), t) = 0$$

We want to solve (I), (II) and (III') for H. To do this we make the claim that

11.5. <u>If there is a germ</u> $\tilde{\xi} : (\mathbf{R}^n \times \mathbf{R}, (0, t_0)) \to \mathbf{R}^n$ <u>with the properties</u>

(a) $$\sum_i \frac{\partial F}{\partial x_i}(x, t) \cdot \tilde{\xi}_i(x, t) + \frac{\partial F}{\partial t}(x, t) = 0$$

(b) $\tilde{\xi}_i(0, t) = 0$ <u>for all</u> i,

<u>then there is also a germ</u> $\tilde{H}$ <u>satisfying</u> (I), (II) <u>and</u> (III').

To prove this one has to solve the differential equation

$$\frac{\partial H}{\partial t}(x, t) = \xi(H(x, t), t)$$

for H with initial condition $H_{t_0} = \text{id}$. The existence of such an H follows from the theory of ordinary differential equations. Substituting H for x in (a) gives (III') and using (b) the equation

$$\frac{\partial H}{\partial t}(0, t) = \xi(H(0, t), t)$$

has the (unique) solution $H(0, t) = 0$. This gives (II) and property (I) is the initial condition above.

Hence we now have to find germs $\tilde{\xi}_i$ which satisfy (a) and (b) above.

We denote the ring of germs $(\mathbf{R}^n \times \mathbf{R}, (0, t_0)) \to \mathbf{R}$, which we are working with, by $\mathcal{E}(n + 1)$. Let $\mathcal{E}(n) \subset \mathcal{E}(n + 1)$ be the ring of those germs which do not depend on t and denote the maximal ideal in $\mathcal{E}(n)$ by $\mathfrak{m}(n)$. We are looking for certain elements $\tilde{\xi}_i$ in $\mathcal{E}(n + 1)$. Condition (b) means that

(b) $\tilde{\xi}_i \in \mathfrak{m}(n) \cdot \mathcal{E}(n + 1)$, for each i.

This implies, by 4.2, that for each i there are $\tilde{\gamma}_j \in \mathcal{E}(n + 1)$, $j = 1, \ldots, n$ such that

$$\xi_i(x, t) = \sum_j x_j \gamma_j(x, t).$$

Hence the existence of $\tilde{\xi}_i$ satisfying (a) and (b) is equivalent to the assertion about germs in $\mathscr{E}(n + 1)$ that

$$\frac{\partial F}{\partial t} \in \mathfrak{m}(n) . \langle \frac{\partial F}{\partial x_1}, \ldots, \frac{\partial F}{\partial x_n} \rangle_{\mathscr{E}(n+1)}$$

However $\frac{\partial F}{\partial t} = \frac{\partial}{\partial t}((1 - t)f + tg) = g - f \in \mathfrak{m}(n)^{k+1}$, using the assumption that f and g have the same k-jet. It is thus sufficient to show that

$$\text{(a, b)} \quad \mathfrak{m}(n)^{k+1} \subset \mathfrak{m}(n) . \langle \frac{\partial F}{\partial x_i} \rangle_{\mathscr{E}(n+1)} .$$

The hypothesis in the theorem gives

$$(*) \qquad \mathfrak{m}(n)^k . \mathscr{E}(n + 1) \subset \mathfrak{m}(n) . \langle \frac{\partial f}{\partial x_i} \rangle_{\mathscr{E}(n+1)} + \mathfrak{m}(n)^{k+1} . \mathscr{E}(n + 1)$$

$$\subset \mathfrak{m}(n)\langle \frac{\partial F}{\partial x_i} \rangle_{\mathscr{E}(n+1)} + \mathfrak{m}(n + 1) . \mathfrak{m}(n)^k . \mathscr{E}(n + 1).$$

For the second inclusion, observe that

$$\frac{\partial F}{\partial x_i} - \frac{\partial f}{\partial x_i} = t . \frac{\partial}{\partial x_i}(g - f) \in \mathfrak{m}(n)^k,$$

and $\mathfrak{m}(n) \subset \mathfrak{m}(n + 1)$.

Now $\mathfrak{m}(n)^k . \mathscr{E}(n + 1)$ is a finitely generated $\mathscr{E}(n+1)$-module - it is generated by monomials in the x_i of degree k. When we apply the Nakayama lemma in the form

$$A \subset B + \mathfrak{m} . A \Rightarrow A \subset B$$

to the inclusion (*), we obtain (a, b):

$$\mathfrak{m}(n)^{k+1} \subset \mathfrak{m}(n)^k . \mathscr{E}(n + 1) \subset \mathfrak{m}(n)\langle \frac{\partial F}{\partial x_i} \rangle_{\mathscr{E}(n+1)} . \quad \checkmark$$

11. 6. Right-transformation and right-equivalence may also be defined for jets, that is, power series, or power series modulo $\hat{\mathfrak{m}}(n)^k$ (i. e. elements in $\mathbf{R}[x_1, \ldots, x_n]/\langle x_1, \ldots, x_n \rangle^k$).

Let

$$\hat{\mathscr{B}}_k(n) = \text{the group of k-jets of germs in } \mathscr{B}(n)$$

$$= \text{the group of n-tuples of polynomials of degree } k\text{:}$$

$$f(x) = (f_1(x_1, \ldots, x_n), \ldots, f_n(x_1, \ldots, x_n))$$

such that $Df(0) = (\partial f_i/\partial x_j(0))$ is invertible, and $f(0) = 0$. The group multiplication is the one induced from the composition of maps.

Let

$$\hat{\mathscr{E}}_k(n, n) = (\mathbf{R}[x_1, \ldots, x_n]/\langle x_1, \ldots, x_n\rangle^{k+1})^n .$$

This is euclidean space of dimension $n \cdot \binom{n+k}{k}$. Since $\hat{\mathcal{B}}_k(n)$ is the subset of $\hat{\mathscr{E}}_k(n, n)$ defined by $\det(\partial f_i/\partial x_j(0)) \neq 0$ and $f(0) = 0$, it is a differentiable manifold.

The group structure:

$$\hat{\mathcal{B}}_k(n) \times \hat{\mathcal{B}}_k(n) \to \hat{\mathcal{B}}_k(n)$$
$$(f(x), g(x)) \mapsto f(g(x)) \text{ modulo } \langle x_1, \ldots, x_n\rangle^{k+1}$$

is a differentiable map. The coefficients of $\hat{f} \circ \hat{g}$ are computed from those of $\hat{f}$ and $\hat{g}$ by applying canonical polynomials. The map

$$\mathcal{B}_k(n) \to \mathcal{B}_k(n)$$
$$\hat{f}(x) \mapsto \hat{f}^{-1}(x) \text{ modulo } \langle x_1, \ldots, x_n\rangle^{k+1}$$

is also differentiable.

Any differentiable manifold with a differentiable group structure is called a Lie group.

11.7. Exercise. Prove that $\dim \hat{\mathscr{E}}_k(n, m) = m. \sum_{i=0}^{k} \binom{n+i-1}{i} = m\binom{n+k}{k}$.

The Lie group $\hat{\mathcal{B}}_k(n)$ acts on $\hat{\mathscr{E}}_k(n, m)$ by composition $\hat{\mathscr{E}}_k(n, m) \times \hat{\mathcal{B}}_k(n) \to \hat{\mathscr{E}}_k(n, m) : (\hat{f}, \hat{g}) \to \hat{f} \circ \hat{g}$.

This action is compatible with first letting $\mathcal{B}(n)$ operate on $\mathscr{E}(n, m)$ and then taking jets. The following diagram commutes:

$$\begin{array}{ccc} \mathscr{E}(n, m) \times \mathcal{B}(n) & \longrightarrow & \mathscr{E}(n, m) \\ \downarrow j^k \times j^k & & \downarrow j^k \\ \hat{\mathscr{E}}_k(n, m) \times \hat{\mathcal{B}}_k(n) & \longrightarrow & \hat{\mathscr{E}}_k(n, m) \end{array} .$$

The right-equivalence classes in $\mathscr{E}(n, m)$ are the orbits of the action of $\mathcal{B}(n)$. The equivalence classes of their k-jets are the orbits of the action of $\hat{\mathcal{B}}_k(n)$ on $\hat{\mathscr{E}}_k(n, m)$. Hence if $f, g \in \mathscr{E}(n)$ are right-equivalent then $\hat{f}$ and $\hat{g}$ lie on the same orbit of the action of $\hat{\mathcal{B}}_k(n)$ on $\hat{\mathscr{E}}_k(n)$. By considering these finite-dimensional orbits in a finite-dimensional euclidean space, we obtain a necessary condition for right-equivalence.

11.8. Lemma. Let $f \in \mathscr{E}(n)$ be a germ with k-jet $\hat{f} \in \hat{\mathscr{E}}_k(n)$. Denote the orbit of $\hat{f}$ under the action $\hat{\mathscr{E}}_k(n) \times \hat{\mathcal{B}}_k(n) \to \hat{\mathscr{E}}_k(n)$ by $\hat{f}\hat{\mathcal{B}}_k(n)$. Let $T_{\hat{f}}\hat{f}\hat{\mathcal{B}}_k(n)$ be the tangent space to this orbit at $\hat{f}$, considered as a subspace of the euclidean space $\hat{\mathscr{E}}_k(n)$. Then

$$T_{\hat{f}}\hat{f}\hat{\mathcal{B}}_k(n) = \mathfrak{m}(n)\langle \frac{\partial f}{\partial x_1}, \ldots, \frac{\partial f}{\partial x_n} \rangle \text{ modulo } \mathfrak{m}(n)^{k+1}.$$

Proof. Consider a germ $\tilde{\delta} : (\mathbf{R}^{n+1}, 0) \to (\mathbf{R}^n, 0)$, $(x, t) \mapsto \delta(x, t) = \delta_t(x)$ with $\tilde{\delta}_0 = \mathrm{id} : (\mathbf{R}^n, 0) \to (\mathbf{R}^n, 0)$.

This is the germ of a differentiable, one-parameter family of transformations in $\mathcal{B}(n)$ which starts with the identity. Such a germ induces a path-germ: $(\mathbf{R}, 0) \to (\hat{\mathscr{E}}_k(n), \hat{f})$, $t \mapsto \hat{f} \circ \hat{\delta}_t$. The 'velocity vectors of these paths at time 0' form the tangent space $T_{\hat{f}}\hat{f}\hat{\mathcal{B}}_k(n)$. If $\delta(t, x)$ is written in the form $x + \varepsilon(t, x)$ then the germ $\tilde{\varepsilon} \in \mathscr{E}(n+1, n)$ is restricted only by the conditions $\varepsilon(0, x) = 0$, $\varepsilon(t, 0) = 0$. The following vectors, when reduced modulo $\mathfrak{m}(n)^{k+1}$, therefore give all the tangent vectors

$$\frac{\partial}{\partial t}(f(x + \varepsilon(t, x)))\Big|_{t=0} = \sum_{i=1}^{n} \frac{\partial f}{\partial x_i} \cdot \frac{\partial \varepsilon_i}{\partial t}\Big|_{t=0}.$$

Since ε has only to fulfil the conditions described above, the derivative $\partial\varepsilon_i/\partial t|_{t=0}$ can be any element in $\mathfrak{m}(n)$. This means that the tangent space is exactly $\mathfrak{m}(n)\langle \partial f/\partial x_i \rangle$ modulo $\mathfrak{m}(n)^{k+1}$. ✓

It was remarked before that determinacy was really a property of jets and now we make this more explicit. Analogously to 11.3, an r-jet $\hat{f} \in \hat{\mathscr{E}}_r(n)$ is called k-determined (for some $k \le r$) if for all $\hat{g} \in \hat{\mathscr{E}}_r(n)$ with $\pi^r_k\hat{f} = \pi^r_k\hat{g} \in \hat{\mathscr{E}}_k(n)$, there is an $\hat{h} \in \hat{\mathcal{B}}_r(n)$ such that $\hat{f} = \hat{g} \circ \hat{h}$.

11.15. Lemma. <u>If</u> $\tilde{f} \in \mathcal{A}(n)$, <u>then</u> $\dim_{\mathbf{R}} \mathcal{E}(n)/\langle \partial f/\partial x_i \rangle_{\mathcal{E}(n)} < \infty$ <u>if and only if</u> $\dim_{\mathbf{C}} \mathcal{O}(n)/\langle \partial f/\partial z_i \rangle_{\mathcal{O}(n)} < \infty$.

Proof. Denote the maximal ideals in $\mathcal{A}(n)$ and $\mathcal{O}(n)$ by $m_{\mathcal{A}}(n)$ and $m_{\mathcal{O}}(n)$ respectively.

'$\Rightarrow$' : $m(n)^k \subset \langle \frac{\partial f}{\partial x_i} \rangle_{\mathcal{E}(n)}$ implies $m(n)^k \subset \langle \frac{\partial f}{\partial x_i} \rangle_{\mathcal{E}(n)} + m(n)^{k+1}$.

We may also deduce that

$$(*) \qquad m_{\mathcal{A}}(n)^k \subset \langle \frac{\partial f}{\partial x_i} \rangle_{\mathcal{A}(n)} + m_{\mathcal{A}}(n)^{k+1}$$

because a monomial of degree k, $\phi \in m_{\mathcal{A}}(n)^k$ can first be written

$$\phi(x) = \sum \lambda_i(x) \frac{\partial f}{\partial x_i}(x) \text{ modulo } m(n)^{k+1},$$

and then each λ_i may be replaced by its k-th Taylor polynomial. The remainder will lie in $m(n)^{k+1}$ but also be analytic, hence (*).

Nakayama's lemma gives

$$m_{\mathcal{A}}(n)^k \subset \langle \frac{\partial f}{\partial x_i} \rangle_{\mathcal{A}(n)}.$$

In particular, each monomial ϕ of degree k belongs to $\langle \partial f/\partial x_i \rangle_{\mathcal{A}(n)}$ which is contained in $\langle \partial f/\partial x_i \rangle_{\mathcal{O}(n)}$. However such monomials generate $m_{\mathcal{O}}(n)^k$ so that

$$m_{\mathcal{O}}(n)^k \subset \langle \frac{\partial f}{\partial z_i} \rangle_{\mathcal{O}(n)}.$$

'$\Leftarrow$' : The derivatives $\partial f/\partial z_i(z)$ are all real when z is taken to be real. That means that when real parts are taken in $m_{\mathcal{O}}(n)^k \subset \langle \partial f/\partial z_i \rangle_{\mathcal{O}(n)}$ we obtain $m_{\mathcal{A}}(n)^k \subset \langle \partial f/\partial z_i \rangle_{\mathcal{A}(n)}$, and hence

$$m_{\mathcal{A}}(n)^k \subset \langle \frac{\partial f}{\partial z_i} \rangle_{\mathcal{E}(n)}. \quad \checkmark$$

The lemma shows that if $\tilde{f}$ is real-analytic and the complex germ $\tilde{f} \in \mathcal{O}(n)$ is algebraically isolated then $\tilde{f}$ is algebraically isolated as a real germ.

11.16. Theorem. <u>A complex-analytic singularity is algebraically isolated if and only if it is isolated.</u>

Proof. Let $\tilde{f} \in \mathcal{O}(n)$ be algebraically isolated, then $m_{\mathcal{O}}(n)^k \subset \langle \partial f/\partial z_i \rangle$. In particular $z_i^k \in \langle \partial f/\partial z_i \rangle$ so that

$$\Sigma(\tilde{f}) \subset \{z \mid Df(z) = 0\}^{\sim} \subset \{z \mid z_i^k = 0\}^{\sim} = \{0\}^{\sim}.$$

For the converse, suppose that $\{0\}^{\sim} = (\Sigma(f))^{\sim}$. This implies $\{0\}^{\sim} = \{z \mid Df(z) = 0\}^{\sim}$ because there would otherwise be a real analytic curve $\phi(t)$ with $\phi(0) = 0$ which was contained in $\{z \mid Df(z) = 0\}$ (curve selection lemma - see Milnor). Along this curve Df would vanish, hence f would also vanish, since $f(0) = 0$. But then the curve would lie in $\Sigma(f)$ and the origin would not be an isolated singularity.

The germ of the set of zeros of the ideal $\langle \partial f/\partial z_i \rangle$ consists just of the origin. The ideal of all germs which vanish on $\{0\}$ is $\langle z_i \rangle$. The Nullstellensatz for holomorphic germs (see Gunning and Rossi) states, for this case, that the second ideal is the radical of the first. The radical is the set of those germs $\tilde{g}$, some power of which lies in $\langle \partial f/\partial z_i \rangle$. Hence $(z_i)^{k_j} \in \langle \partial f/\partial z_i \rangle$ and it follows that $m_{\mathcal{O}}(n)^k \subset \langle \partial f/\partial z_i \rangle$ for large enough k. √

One particular deduction to be made from the preceding theorems is that a holomorphic germ with isolated singularity at the origin can always be transformed into a polynomial using holomorphic coordinate changes. For analytic germs, all the foregoing transformations could have been chosen to be analytic.

The question as to when a germ $\tilde{f} \in \mathcal{E}(n, m)$ is finitely determined for $m > 1$, may be handled in the same way as for $m = 1$. However, for $m > 1$, a germ is finitely determined if and only if it has rank m. For the general case, it is more sensible to include transformations of the range $\mathbf{R}^m$ and to study the orbits of the action of $\mathcal{B}(n) \times \mathcal{B}(m)$ on $\mathcal{E}(n, m)$. There are explicit conditions about inclusions of finite-dimensional vector spaces, similar to those we have discussed. This will be taken no further here. The method of proof is the same as was presented above in Mather's theorem. However where we have used the Nakayama lemma, one has to apply the Malgrange preparation theorem.

12 · Some elementary algebraic geometry

Literature: S. Lang: Algebra, Addison Wesley (1969). Introduction to algebraic geometry, Wiley (1964).
W. V. D. Hodge and D. Pedoe: Methods of algebraic geometry, Cambridge Univ. Press (1954).
H. Whitney: Elementary structure of real algebraic varieties, Ann. of Math., 66 (1957), 545-56.

This chapter puts together a few facts from algebraic geometry which will be used in the next chapter. The material presented would have been too extensive were it not that it also introduces some of the techniques for dissecting algebraic sets. This is a first step towards stratification theory which plays an important part in the study of singularities.

Let K be a field - usually $\mathbf{R}$ or $\mathbf{C}$ - and let K^n be the vector space of n-tuples over K.

12.1. Definition. A subset $A \subset K^n$ is called algebraic if there are polynomials $f_1, \ldots, f_r \in K[x] = K[x_1, \ldots, x_n]$ such that

$$A = \{x \in K^n \mid f_1(x) = \ldots = f_r(x) = 0\}$$

(we shall write $x = (x_1, \ldots, x_n)$).

If $A \subset K^n$ is an arbitrary subset, then the ideal of all polynomials vanishing on A will be denoted $\mathfrak{n}(A)$. Hence

$$\mathfrak{n}(A) = \{f \in K[x] \mid f(a) = 0 \text{ for all } a \in A\}.$$

Conversely an ideal $\mathfrak{a} \subset K[x]$ defines a subset $V(\mathfrak{a}) \subset K^n$. This is the set of zeros of $\mathfrak{a}$:

$$V(\mathfrak{a}) = \{x \in K^n \mid f(x) = 0 \text{ for all } f \in \mathfrak{a}\}.$$

The subset $V(\mathfrak{a})$ is algebraic because of the following result:

12. 2. Hilbert's basis theorem. The polynomial ring $K[x]$ is Noetherian, that is, every ideal is finitely generated (see Lang: Algebra). √

Hence $V(\mathfrak{a})$ is the set of zeros of the generators $\{f_1, \ldots, f_r\}$ of $\mathfrak{a}$. By definition, a set A is algebraic if $V(\mathfrak{n}(A)) = A$. For any subset, $V(\mathfrak{n}(A)) \supset A$. Hence, in general, $V(\mathfrak{n}(A))$ is the smallest algebraic subset containing A.

If A and B are algebraic sets then,

$$A \subset B \Longleftrightarrow \mathfrak{n}(A) \supset \mathfrak{n}(B)$$

and for ideals $\mathfrak{a}$ and $\mathfrak{b} \subset K[x]$:

$$\mathfrak{a} \subset \mathfrak{b} \Longleftrightarrow V(\mathfrak{a}) \supset V(\mathfrak{b}).$$

To any decreasing sequence of algebraic sets

$$A_1 \supset A_2 \supset A_3 \supset \ldots$$

there corresponds the increasing sequence of ideals which vanish on them,

$$\mathfrak{a}_1 \subset \mathfrak{a}_2 \subset \mathfrak{a}_3 \subset \ldots \qquad (\mathfrak{a}_i = \mathfrak{n}(A_i)) .$$

Because $K[x]$ is Noetherian, there is an n such that all the generators of the ideal $\mathfrak{a} = \bigcup_{i=1}^{\infty} \mathfrak{a}_i$ are contained in $\mathfrak{a}_n$, hence $\mathfrak{a} = \mathfrak{a}_n$. The sequence of ideals stops increasing after n and therefore the sequence of algebraic sets stops decreasing after n. Thus

12. 3. Every strictly decreasing sequence of algebraic sets is finite (basis theorem).

It follows that one may give K^n a topology in which the algebraic sets form the closed sets. This is the Zariski topology and is much weaker than the usual one when $K = \mathbf{R}$ or $\mathbf{C}$.

The union of two algebraic sets is obviously algebraic since $V(\mathfrak{a}) \cup V(\mathfrak{b}) = V(\mathfrak{a} \cap \mathfrak{b})$. Any intersection of algebraic sets is in fact finite, as we have seen, and the set of polynomials defining $A \cap B$ is the union of the sets of defining polynomials for A and for B.

Any algebraic subset has a topology, induced from the Zariski topology on K^n, and for any subset $A \subset K^n$, the set $V(\mathfrak{n}(A))$ is the Zariski-closure of A.

From the definitions $\mathfrak{n}(V(\mathfrak{a}))$ always contains $\mathfrak{a}$, but equality does not hold in general.

12. 4. Nullstellensatz (Hilbert). _Let_ K _be algebraically closed, then_ $\mathfrak{n}(V(\mathfrak{a}))$ _is the radical of_ $\mathfrak{a}$, _that is_

$$\mathfrak{n}(V(\mathfrak{a})) = \{f \in K[x] \mid f^r \in \mathfrak{a} \text{ for some } r\}$$

(see Lang: Algebra). ✓

It is obvious that the radical of $\mathfrak{a}$ is contained in $\mathfrak{n}(V(\mathfrak{a}))$, since

$$f^r \in \mathfrak{a} \Rightarrow f^r | V(\mathfrak{a}) = 0 \Rightarrow f | V(\mathfrak{a}) = 0 \Rightarrow f \in \mathfrak{n}(V(\mathfrak{a})).$$

12. 5. Definition. An algebraic set A is called _irreducible_ or a _variety_ if whenever A_1 and A_2 are algebraic and $A = A_1 \cup A_2$, then $A = A_1$ or $A = A_2$.

Thus, a variety cannot be decomposed into smaller algebraic sets. By taking complements in the Zariski topology, the definition may be rephrased: an algebraic subset A is a variety if the intersection of non-empty, open subsets of A is non-empty.

An arbitrary algebraic set A, which is not irreducible, can be decomposed into algebraic sets: $A = A_1 \cup A_2$. If this process is iterated then one finally arrives at a decomposition $A = A_1 \cup \ldots \cup A_r$ into irreducible sets (basis theorem). This decomposition is unique up to a change in the order of its members so long as there are no inclusions $A_i \subset A_j$ for $i \neq j$. For, if $A = B_1 \cup \ldots \cup B_s$ is a second decomposition without inclusions, then for each i, there exist j and k with $A_i \subset B_j \subset A_k$ because of the irreducibility of A_i and B_j. Since there are no inclusions, $i = k$, and $A_i = B_j$.

The irreducible sets in the decomposition of A above are called the irreducible components of A.

12.6. An algebraic set V is irreducible if and only if the ideal $\mathfrak{n}(V)$ is prime.

($\mathfrak{n}$ is prime when $f.g \in \mathfrak{n} \Rightarrow f \in \mathfrak{n}$ or $g \in \mathfrak{n}$.)

Proof. If V is irreducible and $f.g \in \mathfrak{n}(V)$, then $V \subset V(\langle f \rangle) \cup V(\langle g \rangle)$ and so, without loss, $V \subset V(\langle f \rangle)$, hence $f \in \mathfrak{n}(V)$.

Conversely suppose V were reducible, $V \subset A \cup B$, $V \not\subset A$ and $V \not\subset B$. Choose $f \in \mathfrak{n}(A)$, $f \notin \mathfrak{n}(B)$ and $g \in \mathfrak{n}(B)$, $g \notin \mathfrak{n}(A)$. Then

$$f.g \in \mathfrak{n}(A \cup B) \subset \mathfrak{n}(V) \text{ but } f \notin \mathfrak{n}(V) \text{ and } g \notin \mathfrak{n}(V). \quad \checkmark$$

If $\mathfrak{a} \subset K[x]$ is an ideal, then the rank of $\mathfrak{a}$ is defined to be

$$\rho(\mathfrak{a}) = \max_{x \in V(\mathfrak{a})} Rk_x(f_1, \ldots, f_k)$$

where $f_1, \ldots, f_k$ is any system of generators for $\mathfrak{a}$, and

$$Rk_x(f_1, \ldots, f_k) = \text{Rank}\left(\frac{\partial f_i}{\partial x_j}(x)\right).$$

The rank ρ does not depend on the choice of generators because if $g_1, \ldots, g_m$ is another system of generators then $g_i = \sum_j a_{ij} f_j$ for $a_{ij} \in K[x]$. Since $f_i(x) = 0$ for $x \in V$ we obtain

$$\left(\frac{\partial g_i}{\partial x_r}\right) = (a_{ij})\left(\frac{\partial f_j}{\partial x_r}\right) \text{ on } V.$$

Hence $Rk_x(g_1, \ldots, g_m) \leq Rk_x(f_1, \ldots, f_k)$ for $x \in V$ and the corresponding result in the other direction gives equality.

Now let $V \neq \emptyset$ be a variety (irreducible) and for simplicity take $K = \mathbf{R}$ or $\mathbf{C}$ from now on. Since $\mathfrak{n}(V)$ is a prime ideal, $K[x]/\mathfrak{n}(V)$ is an integral domain. We consider the quotient field:

$$K(V) = Q(K[x]/\mathfrak{n}(V)).$$

The vector space $(K(V))^n$ of n-tuples over the field $K(V)$ contains the point $x = (x_1, \ldots, x_n)$ whose coordinates are the indeterminates x_i

modulo $\mathfrak{n}(V)$. This point is called the generic point of V. Now there is a canonical inclusion $K \subset K(V)$ and so we can substitute points of $(K(V))^n$ into any polynomial $f \in K[x]$. By definition of the generic point, we obtain:

12.7. A polynomial $f \in K[x]$ vanishes on V if and only if f vanishes at the generic point:

$$f \in \mathfrak{n}(V) \Longleftrightarrow f(x) = 0 \text{ in } K(V).$$

In particular the rank of $\mathfrak{n}(V)$ is equal to the rank of a system of generators $(f_1, \ldots, f_k)$ for $\mathfrak{n}(V)$ at the generic point. This is because any minor Φ in the matrix $(\partial f_i/\partial x_j)$ satisfies $\Phi(V) = 0$ if and only if $\Phi(x) = 0$ for the generic point $x \in K(V)$.

12.8. **Definition.** The dimension of a variety V is the transcendence degree of $K(V)$ over K.

12.9. **Theorem.** The dimension of a variety $V \subset K^n$ is equal to the corank of $\mathfrak{n}(V)$, that is, $\dim V = n - \rho(\mathfrak{n}(V))$.

Proof. Let $\dim(V) = d$ = transcendence degree of $K(V)$ over K. With a suitable numbering of the coordinates, the elements $x_{d+i} \in K(V)$ are algebraic over the field $K(x_1, \ldots, x_d)$. Hence there is an irreducible polynomial g_i (a polynomial of smallest degree) such that $g_i(x_1, \ldots, x_d, x_{d+i}) = 0$ in $K(V)$ or, equivalently, $g_i \in \mathfrak{n}(V)$. In particular $\partial g_i/\partial x_{d+i}(x) \neq 0$ in $K(V)$. (If $\partial g_i/\partial x_{d+i} \equiv 0$, then g_i is constant as a polynomial in the indeterminate x_{d+i}. Then $(x_1, \ldots, x_d)$ would be algebraically dependent. If $\partial g_i/\partial x_{d+i} \not\equiv 0$, then $\partial g_i/\partial x_{d+i} = 0$ would be an algebraic relation for x_d of lower degree than $g_i(x) = 0$.)

Hence, it follows that the matrix $(\partial g_i/\partial x_j(x))$ has rank $n - d$ at the generic point and $g_i \in \mathfrak{n}(V)$, for all i. Thus $\rho(\mathfrak{n}(V)) \geq n - d$.

Now we show that $\rho \leq n - d$.

To do this we use two facts. First, derivations D_i of $K(x_1, \ldots, x_n)$ with the property $D_i|K = 0$ are uniquely defined by the equation $D_i x_j = \delta_{ij}$. Secondly, any derivation of a field (in this case $K(x_1, \ldots, x_d)$) may be extended uniquely to an algebraic extension of the field (here K(V)).

See Lang: Algebra, X, §7, Prop. 10. The definition of D_i on the rational functions in K(x) uses the usual rules of differential calculus. The extension of D from K to K(y) for an algebraic element y is determined by the minimal polynomial for y. If $p(y) = \sum p_j y^j$ is the minimal polynomial and $p^D(y)$ is defined as $\sum D(p_j)y^j$, then

$$D(p(y)) = 0 = p^D(y) + p'(y). D(y).$$

Since p is minimal, $p'(y) \neq 0$ and we have a definition for D(y). This defines a derivation on K(y).

The vector space of derivations on K(V) which vanish on K therefore has dimension d in our case. A basis of this space is $\{D_i \mid i = 1, \ldots, d\}$ where $D_i(x_j) = \delta_{ij}$ for $j \leq d$. If $\{f_1, \ldots, f_k\}$ is a system of generators for $\mathfrak{n}(V)$, then at the generic point x we have $f_j(x) = 0$. Hence

$$\frac{\partial f_j}{\partial x_i} + \sum_{\nu=1}^{n-d} \frac{\partial f_j}{\partial x_{d+\nu}} . D_i(x_{d+\nu}) = 0, \qquad i = 1, \ldots, d.$$

Also, for the minimal polynomials g_ν of $x_{d+\nu}$ we have

$$\frac{\partial g_\nu}{\partial x_i} + \frac{\partial g_\nu}{\partial x_{d+\nu}} . D_i(x_{d+\nu}) = 0, \qquad \nu = 1, \ldots, n-d.$$

By substituting from these equations into the preceding ones, we find that at the generic point

$$\frac{\partial f_j}{\partial x_i} = \sum_{\nu=1}^{n-d} \frac{\partial f_j}{\partial x_{d+\nu}} . \frac{\partial g_\nu}{\partial x_i} . \left(\frac{\partial g_\nu}{\partial x_{d+\nu}}\right)^{-1} \text{ for } i = 1, \ldots, d.$$

However this shows that in the matrix Df(x), $x \in K(V)^n$, the first d columns are linear combinations of the last n - d. Hence $\rho \leq n-d$. √

12.10. Theorem. Let $\mathfrak{a} \subset \mathfrak{b}$ be prime ideals in K[x] and let $d(\mathfrak{a})$ be the transcendence degree of $K(\mathfrak{a})$ over K, where $K(\mathfrak{a}) = Q(K[x]/\mathfrak{a})$. Define $d(\mathfrak{b})$ similarly. Then $d(\mathfrak{a}) \geq d(\mathfrak{b})$ and if $\mathfrak{a} \neq \mathfrak{b}$, then $d(\mathfrak{a}) > d(\mathfrak{b})$.

Proof. Suppose $x_1, \ldots, x_d$ are algebraically independent in $K(\mathfrak{b})$ with an algebraic relation $f(x_1, \ldots, x_d) = 0$ in $K(\mathfrak{a})$. This means $f(x_1, \ldots, x_d)$ belongs to $\mathfrak{a}$ but not to $\mathfrak{b}$ which contradicts the hypothesis. Hence $d(\mathfrak{a}) \geq d(\mathfrak{b})$.

Assume now that $d(\mathfrak{a}) = d(\mathfrak{b})$, and that $\{x_1, \ldots, x_d\}$ is a transcendence basis for $K(\mathfrak{b})$. Suppose $f \in \mathfrak{b}$, then we shall show that $f \in \mathfrak{a}$. Now f represents an element in $K(\mathfrak{a})$ which is algebraic over $K(x_1, \ldots, x_d)$. Hence there is a polynomial g in $K(x_1, \ldots, x_d)[t]$ such that

$$g(x_1, \ldots, x_d, f) = 0 \text{ in } K(\mathfrak{a}).$$

We multiply this by the product $q(x_1, \ldots, x_d) \neq 0$ of the denominators of the coefficients of g in $K(x_1, \ldots, x_d)$. This gives

$$h(x, t) = q(x).g(x, t) \in K[x_1, \ldots, x_d, t].$$

This satisfies $h(x_1, \ldots, x_d, f) = 0$ in $K(\mathfrak{a})$ and so $h(x_1, \ldots, x_d, f) \in \mathfrak{a}$. We can assume that $h(x_1, \ldots, x_d, 0) \neq 0$ in $K(\mathfrak{a})$ by choosing g to be minimal with respect to t, unless $f = 0$ in $K(\mathfrak{a})$, i.e. $f \in \mathfrak{a}$.

Now consider the projection

$$\phi : K[x]/\mathfrak{a} \to K[x]/\mathfrak{b},$$

$\phi(h(x_1, \ldots, x_d, f))$ is equal to $h(x_1, \ldots, x_d, 0)$ since $\phi(f) = 0$, where these polynomials are understood modulo $\mathfrak{a}$ or $\mathfrak{b}$ as appropriate. From the construction $h(x_1, \ldots, x_d, f) = 0$ in $K[x]/\mathfrak{a}$ and so $h(x_1, \ldots, x_d, 0) = 0$ in $K(\mathfrak{b})$. It follows that $h(x_1, \ldots, x_d, 0)$ is the zero polynomial because $x_1, \ldots, x_d$ are algebraically independent. This contradicts the earlier assumption unless $f \in \mathfrak{a}$ as we wanted to show. ✓

In particular if we choose $\mathfrak{a} = \mathfrak{n}(W)$ and $\mathfrak{b} = \mathfrak{n}(V)$ we obtain:

12.11. Corollary. If $V \subset W$ are varieties, then $\dim V \leq \dim W$ and $\dim V = \dim W$ if and only if $V = W$. ✓

For an arbitrary algebraic set A the dimension of A, $\dim A$, is defined to be the largest of the dimensions of irreducible components of A, and $(n - \dim A)$ is the codimension of A. Hence in general, for algebraic sets, $A \subset B$ implies $\dim A \leq \dim B$. If B is irreducible, then $A \subset B$ and $\dim A = \dim B$ imply $A = B$.

If A is an algebraic set, defined by the polynomials $\{f_1, \ldots, f_k\}$, then the singular locus of A is given by

$$\Sigma A = \{x \in A \,|\, Rk_x(f_1, \ldots, f_k) \text{ is not maximal on } A\}.$$

The set $A - \Sigma A$ is called the regular locus of A, its points are called regular. Thus the regular points of A are those x, where $Rk_x(f_1, \ldots, f_k) = \rho(\mathfrak{n}(A)) = n - \dim(A)$. By definition of the rank of an ideal, the regular locus of A is not empty. Hence ΣA is strictly smaller than A. Further, ΣA is an algebraic set, defined as a subset of A, by the vanishing of all $(\rho \times \rho)$-minors in the matrix $(\partial f_i / \partial x_j)$.

The connection between the definition of dimension of an algebraic set given above and the topological definition of dimension is explained in the next theorem.

12.12. **Theorem.** If $V \subset K^n$ is a variety, then the regular locus $V - \Sigma V$ is an analytic manifold (real or complex according to context) of dimension $\dim V$.

Assume this result for the moment and decompose the algebraic set A into irreducible subsets, $A = V_1 \cup \ldots \cup V_r$. The set

$$A - (\bigcup_{i \neq j} (V_i \cap V_j) \cup \bigcup_i \Sigma V_i)$$

can be decomposed into the union of the r non-empty, analytic manifolds:

$$V_i - ((V_i \cap \bigcup_{j \neq i} V_j) \cup \Sigma V_i).$$

The subsets which have been removed are algebraic and so the procedure may be iterated. Using the algebraic definition, the dimension of $(V_i \cap \bigcup_{j \neq i} V_j) \cup \Sigma V_i$ is smaller than that of V_i so that the iteration will stop after $(\dim A)$ steps. The construction decomposes A into a disjoint union of analytic manifolds with dimensions less than or equal to $\dim A$ (one at least has the same dimension as A).

Proof of the theorem. Let $x \in V - \Sigma V$ and $f_1, \ldots, f_\rho \in \mathfrak{n}(V)$ be polynomials such that $\mathrm{Rk}_x(f_1, \ldots, f_\rho) = \rho = n - \dim V$. Let W be the set $\{x \mid f_1(x) = \ldots = f_\rho(x) = 0\}$, so that $V \subset W$. Because V is irreducible, it is contained in an irreducible component W_0 of W and we have $\rho = \mathrm{Rk}\, \mathfrak{n}(V) \geq \mathrm{Rk}\, \mathfrak{n}(W_0) \geq \rho$. The first inequality here follows from theorem 12.10 and the second because $f_1, \ldots, f_\rho$ belong to $\mathfrak{n}(W_0)$ and $x \in V \subset W_0$. Thus $\mathrm{Rk}\, \mathfrak{n}(V)$ is equal to $\mathrm{Rk}\, \mathfrak{n}(W_0)$ and by Theorem 12.10, $V = W_0$. Therefore V is an irreducible component of W and what we now have to show is that the germ of W at x is irreducible (i.e. locally $W = W_0$). This would complete the proof because the germ of W at x is the solution set of a regular system of ρ polynomial equations and hence an analytic manifold.

To see that there is always a neighbourhood of x in W contained in an irreducible component, observe that it is possible to transform the germ of W at x, analytically, into the germ of $\{x_1 = \ldots = x_k = 0\}$ at the origin. This germ is irreducible as the germ of an analytic set because the ideal of analytic germs which vanish on it is generated by $\{x_1, \ldots, x_k\}$, and this ideal is prime. (Note that if a set were algebraically reducible it would certainly be analytically reducible. Note also that the result corresponding to 12.6 is true for sets defined by analytic equations.) $\checkmark$

As a preliminary to the next theorem we have:

12.13. Lemma. <u>The projection $\pi : K^{n+1} \to K^n$, $(x_1, \ldots, x_{n+1}) \mapsto (x_1, \ldots, x_n)$ is open in the Zariski topology.</u>

Proof. Let $U \subset K^{n+1}$ be open, so that $A = K^{n+1} - U$ is algebraic. Then

$$x \notin \pi(U) \Longleftrightarrow \pi^{-1}(x) \cap U = \emptyset \Longleftrightarrow \pi^{-1}(x) \subset A.$$

Thus we have to show that $V = \{x \in K^n \mid \pi^{-1}(x) \subset A\}$ is algebraic. Suppose that $\mathfrak{n}(A)$ is generated by $\{f_1, \ldots, f_k\}$, $f_i \in K[x_1, \ldots, x_{n+1}]$, and

$$f_i(x_1, \ldots, x_{n+1}) = \sum_j a_{ij}(x_1, \ldots, x_n) . x_{n+1}^j, \quad a_{ij} \in K[x_1, \ldots, x_n].$$

We have: $(x_1, \ldots, x_n) \in V$ if and only if $f(x_1, \ldots, x_n, x_{n+1}) = 0$ for all x_{n+1}, that is, if and only if $a_{ij}(x_1, \ldots, x_n) = 0$. Hence

$$V = \{(x_1, \ldots, x_n) \mid a_{ij}(x_1, \ldots, x_n) = 0 \text{ for all } (i, j)\}. \quad \checkmark$$

12.14. Theorem. Let $V \subset K^n$ be irreducible, then $\pi^{-1}(V) \subset K^{n+1}$ is also irreducible and $\operatorname{codim} V = \operatorname{codim} \pi^{-1}(V)$.

Proof. If $V = \{a\}$ is a point then $\pi^{-1}\{a\}$ is irreducible. This is because the ideal $\langle (x_i - a_i) \mid i = 1, \ldots, n \rangle$, which vanishes on $\pi^{-1}\{a\}$, is the kernel of the map

$$\begin{aligned} K[x_1, \ldots, x_{n+1}] &\to K[x_{n+1}] \\ x_i &\mapsto a_i \quad \text{for } i \le n \\ x_{n+1} &\mapsto x_{n+1}. \end{aligned}$$

$K[x_{n+1}]$ is an integral domain and so the kernel of the map is prime.

In general we must show that if $U_1, U_2 \subset \pi^{-1}(V)$ are open and non-empty, then $U_1 \cap U_2 \neq \emptyset$. Now if $U_i \neq \emptyset$ and open then the same is true of $\pi(U_i)$. Because V is irreducible there must be an element $a \in \pi(U_1) \cap \pi(U_2)$. This gives $\pi^{-1}\{a\} \cap U_i \neq \emptyset$. Now because $\pi^{-1}\{a\}$ is irreducible, the sets $\pi^{-1}\{a\} \cap U_1$ and $\pi^{-1}\{a\} \cap U_2$ have a common point and so $U_1 \cap U_2 \neq \emptyset$.

It is clear that $\mathfrak{n}(V) \subset \mathfrak{n}(\pi^{-1}(V))$ so that $\rho(\mathfrak{n}(V)) \le \rho(\mathfrak{n}(\pi^{-1}(V)))$. Conversely if $f_1, \ldots, f_k \in \mathfrak{n}(\pi^{-1}(V))$ then $f_i(x_1, \ldots, x_n, a) \in \mathfrak{n}(V)$ for any constant $a \in K$. Moreover $\partial f_i / \partial x_{n+1}(x_1, \ldots, x_n, a) = 0$ for $(x_1, \ldots, x_n) \in V$. If one chooses a point $(x_1, \ldots, x_n, a) \in \pi^{-1}(V)$, then

$$\operatorname{Rk}_{(x, a)}(f_1, \ldots, f_k) = \operatorname{Rk}_x(f_1(x, a), \ldots, f_k(x, a))$$

and $f_i(x, a) \in K[x_1, \ldots, x_n]$, so $\rho(\mathfrak{n}(\pi^{-1}(V))) \le \rho(\mathfrak{n}(V))$. $\checkmark$

The last argument also gives $\pi^{-1}(V - \Sigma V) = \pi^{-1}(V) - \Sigma\pi^{-1}(V)$, which is what one intuitively expects when dimension is interpreted topologically.

12.15. Only example of the chapter.

$V = \{(x, y) \in \mathbf{R}^2 \mid y^2 - x^2(1 - x^2) = 0\}$

$\Sigma V = \{0\}$

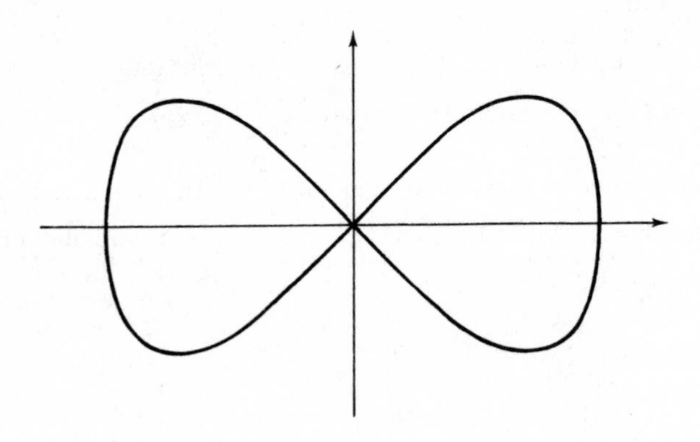

13 · Tougeron's theory

Literature: J. C. Tougeron: Idéaux de fonctions différentiables I, Ann. Inst. Fourier, 18 (1968), 177-240.

The statements we have made about germs have often assumed that a germ $\tilde{f} : (\mathbf{R}^n, 0) \to (\mathbf{R}^n, 0)$ is finite, that is, $\mathscr{E}(n)/\langle f_1, \ldots, f_p\rangle$ has finite dimension. How large is the (unpleasant) set of those germs $\tilde{f}$ for which $\dim \mathscr{E}(n)/\langle f_1, \ldots, f_p\rangle = \infty$?

Consider the implications

(I) $\dim \mathscr{E}(n)/\langle f_1, \ldots, f_p\rangle = k \Rightarrow$

(II) $\dim \mathscr{E}(n)/(\langle f_1, \ldots, f_p\rangle + \mathfrak{m}(n)^{k+1})$

$= \dim \hat{\mathscr{E}}_k(n)/\langle \hat{f}_1, \ldots, \hat{f}_p\rangle \le k \Rightarrow$

(Nakayama's lemma)

$$\mathfrak{m}(n)^k \subset \langle f_1, \ldots, f_p\rangle_{\mathscr{E}(n)} + \mathfrak{m}(n)^{k+1} \Rightarrow$$

(Nakayama's lemma)

$$\mathfrak{m}(n)^{k+1} \subset \mathfrak{m}(n)^k \subset \langle f_1, \ldots, f_p\rangle_{\mathscr{E}(n)}.$$

The last condition and condition (II) together imply $\dim \mathscr{E}(n)/\langle f_1, \ldots, f_p\rangle \le k$. This shows, in particular, that:

13.1. **Remark.** $\dim \mathscr{E}(n)/\langle f_1, \ldots, f_p\rangle < \infty$ if and only if $\dim \hat{\mathscr{E}}_k(n)/\langle \hat{f}_1, \ldots, \hat{f}_p\rangle \le k$ for some k.

This condition is useful because it concerns the sequence of finite dimensional jet spaces. Next we introduce a few new concepts in order that the rest of the chapter may be explained more elegantly. Consider the following sequence of euclidean spaces and projections.

$$\hat{\mathscr{E}}(n, p) \xrightarrow{\pi^\infty} \dots \to \hat{\mathscr{E}}_{k+1}(n, p) \xrightarrow{\pi_k^{k+1}} \hat{\mathscr{E}}_k(n, p) \xrightarrow{\pi_{k-1}^k} \hat{\mathscr{E}}_{k-1}(n, p) \to \dots$$

13.2. **Definition.** A subset $A \subset \hat{\mathscr{E}}(n, p)$ is called <u>proalgebraic</u> if there are algebraic sets $A_k \subset \hat{\mathscr{E}}_k(n, p)$ such that

$$A = \bigcap_{k=1}^{\infty} (\pi_k^\infty)^{-1}(A_k)$$

It is easily seen that this definition is not altered by demanding that $\pi_k^{k+1}(A_{k+1}) \subset A_k$; in this case the <u>codimension</u> of A is the supremum of the codimensions of the A_k. The map π_k^∞ is the canonical projection,

$$\pi_k^\infty : \hat{\mathscr{E}}(n, p) \to \hat{\mathscr{E}}_k(n, p) = \mathscr{E}(n, p)/m(n, p)^{k+1} .$$

13.3. **Exercise.** If the $\hat{\mathscr{E}}_k(n, p)$ are given the Zariski topology, show that the proalgebraic sets are the closed sets of the weakest topology on $\hat{\mathscr{E}}(n, p)$ with the property that all the projections π_k^∞ are continuous.

Having introduced these new ideas, we return to the non-finite germs. Define Y_k as follows

$$Y_k = \{\hat{f} = (\hat{f}_1, \dots, \hat{f}_p) \mid \dim \hat{\mathscr{E}}_k(n)/\langle\hat{f}_1, \dots, \hat{f}_p\rangle > k\}.$$

The set Y_k is a subset of $\hat{\mathscr{E}}_k(n, p)$ and remark 13.1 states that f is non-finite if and only if $j^k(f) \in Y_k$ for all k. If we denote $\bigcap_{k=1}^{\infty} (\pi_k^\infty)^{-1} Y_k$ by Y, then the last condition is equivalent to $j(f) \in Y$.
Notice that $Y_{k+1} \subset \pi^{-1} Y_k$, for if $f \notin \pi^{-1} Y_k$, then $\dim \mathscr{E}(n)/\langle f_1, \dots, f_p\rangle \leq k$ and hence $f \notin Y_{k+1}$ (here $\pi = \pi_k^{k+1}$).

The significant result in this chapter is:

13.4. **Theorem (Tougeron).** <u>The sets Y_k are algebraic. Furthe</u>r <u>Y is a proalgebraic set of infinite codimension when</u> $n \leq p$.

Proof of the first statement. $\hat{f} \in Y_k \iff \dim(\hat{\mathscr{E}}_k(n)/\langle\hat{f}_1, \dots, \hat{f}_p\rangle) > k$
$\iff \dim(\langle\hat{f}_1, \dots, \hat{f}_p\rangle \cdot \hat{\mathscr{E}}_k(n)) < \dim \hat{\mathscr{E}}_k(n) - k$.

Denote $(\dim \hat{\mathscr{E}}_k(n) - k)$ by $r(k)$.

Let $\{\phi_j\}$ be all the monomials of degree $\leq k$ in $\mathscr{E}(n)$, then $(\hat{f}_1, \dots, \hat{f}_p) \in Y_k$ if and only if the linear map

$$\begin{aligned} \mathbf{R}^p \otimes \langle \phi_j \rangle_{\mathbf{R}} &\to \hat{\mathscr{E}}_k(n) \\ e_i \otimes \phi_j &\mapsto j^k(\hat{f}_i \cdot \phi_j) \end{aligned}$$

has rank smaller than $r(k)$. This condition is determined by the vanishing of certain determinants which are polynomials in the coefficients of the k-jets of the f_i. √

To prove the second part we need the following lemma.

13.5. Lemma. Codim $Y = \infty$ if and only if for each k-jet $\hat{f}_k$ there exists an l-jet $\hat{f}_l$ for some $l > k$, such that $\pi^l_k \hat{f}_l = \hat{f}_k$ and $\hat{f}_l \notin Y_l$.

The second condition just says that over any jet there is a finite one.

Proof of lemma. Assume that the supremum of the codimensions of the Y_k is infinite and that no finite jet lies over $\hat{f}_k \in \hat{\mathscr{E}}_k(n)$. Then $(\pi^l_k)^{-1}\hat{f}_k \subset Y_l$, for all $l > k$, and Y_l has at most the same codimension as $(\pi^l_k)^{-1}\hat{f}_k$. But this cannot exceed $\dim(\hat{\mathscr{E}}_k(n))$ and this bound does not depend on l.

Conversely suppose that over each k-jet there is a finite jet. Let $d_i = \operatorname{codim} Y_i$, then

$$d_k \leq d_{k+1} \leq d_{k+2} \leq \dots .$$

The proof is finished unless eventually (or, without loss, straight away)

$$d_k = d_{k+1} = d_{k+2} = \dots .$$

In this case consider an irreducible component X_k of Y_k with highest dimension. We know that $(\pi^l_k)^{-1}(X_k)$ is irreducible for $l > k$ and so for every $l > k$, either

$$(*) \qquad Y_l \cap (\pi^l_k)^{-1}(X_k) = (\pi^l_k)^{-1}(X_k)$$

or the left hand side has higher codimension (use the remark after 12.11).

Denote by b_k the number of irreducible components of Y_k with highest dimension. We argue that $b_k \geq b_{k+1} \geq b_{k+2} \geq \dots$. Any irreducible component of Y_{k+1} is contained in the lift of Y_k under $(\pi_k^{k+1})^{-1}$ and so in the lift of an irreducible component of Y_k. Since $d_k = d_{k+1}$, an irreducible component of Y_{k+1} with highest dimension is contained in, and hence equal to, $(\pi_k^{k+1})^{-1}(X_k)$, where X_k is a similar component of Y_k (as above).

Now b_k is finite, hence eventually $b_k = b_{k+1} = b_{k+2} = \dots$. This means that (*) is always true - no irreducible components of highest dimension are lost. It follows that $(\pi_k^l)^{-1}(X_k) \subset Y_l$ for every l. Therefore if $\hat{f} \in X_k$, then for every l, the l-jet $\hat{f}_l \in (\pi_k^l)^{-1}\{\hat{f}\}$ lies in Y_l. This is a contradiction. ✓

Proof of the second part of the theorem. It is sufficient to show that if f is a p-tuple of polynomials of degree k, then there is a p-tuple of homogeneous polynomials h of degree $k + 1$ so that $(f + h) \notin Y_l$ for some $l > k$.

Choose $g = (x_1^{k+1}, \dots, x_n^{k+1}, 0, \dots, 0)$ (recall $n \leq p$) and put $h_t = (1 - t)f + tg$. The set

$$A = \{t \in \mathbf{R} \mid \hat{h}_t \in Y_l\} \subset \mathbf{R}$$

is algebraic because Y_l is algebraic and the map

$$t \mapsto (1 - t)\hat{f} + t\hat{g}$$

sends t linearly into the coefficients of $\hat{h}_t$ (and linear maps are polynomial maps). Thus A is either finite or all of $\mathbf{R}$.

Now, $1 \notin A$ for large enough l, since $\hat{h}_1 = \hat{g}$ and $\mathcal{E}(n)/\langle x_1^{k+1}, \dots, x_n^{k+1}\rangle$ has finite dimension.

Hence A is a finite set for large enough l. Choose a constant $t \notin A$, $t \neq 1$, then $\hat{h}_t = (1 - t)\hat{f} + t\hat{g} \notin Y_l$ for some l. Because the components of h_t and $h_t/(1 - t)$ generate the same ideal,

$$\left(\frac{1}{1 - t}\right) \cdot \hat{h}_t = \hat{f} + \left(\frac{t}{1 - t}\right) \cdot \hat{g} \notin Y_l$$

for some l, as required. ✓

13.6. Remark. Those $f \in \mathcal{E}(n)$, for which Df is not finite, also form a subset with infinite codimension. To prove this one repeats all the arguments and concludes the proof by observing that $g = (x_1^{k+1}, \ldots, x_n^{k+1})$ may be written $g = Dq$, where $q = (x_1^{k+2} + \ldots + x_n^{k+2})/(k+2)$.

14 · The universal unfolding of a singularity

Literature: G. Wassermann: Stability of unfoldings, Dissertation, Regensburg 1973, Springer Lecture Notes, 393 (1974).

J. Mather: Right equivalence, manuscript.

In order to study a singularity $\eta \in m(n)^2$, that is, a germ $\eta : (\mathbf{R}^n, 0) \to (\mathbf{R}, 0)$, $D\eta(0) = 0$, one imbeds the germ in an r-parameter family of germs in the following way. Let $\mathbf{R}^n \subset \mathbf{R}^{n+r}$ be the subspace where the last r coordinates vanish. Denote a point of $\mathbf{R}^{n+r}$ by $(x, u) = (x_1, \ldots, x_n, u_1, \ldots, u_r)$, $x \in \mathbf{R}^n$, $u \in \mathbf{R}^r$.

14.1. Definition. Let $\eta \in m(n)$ be a singularity. An (r-parameter) unfolding or deformation of η is a germ $\tilde{f} \in m(n+r)$ such that $\tilde{f}|\mathbf{R}^n = \eta$. This unfolding will be denoted $(r, \tilde{f})$.

If f is a representative of $\tilde{f}$ and (x_0, u_0) a point near the origin, then $f|(\mathbf{R}^n \times \{u_0\}, (x_0, u_0))$ defines a germ which is near η.

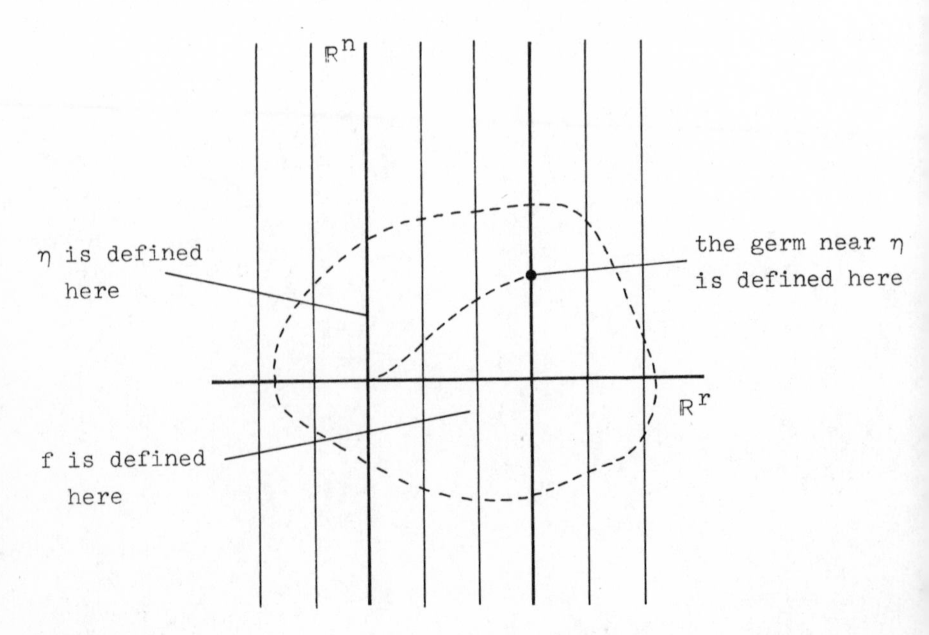

If one follows a path from the origin to the point (x_0, u_0), then along the path the germ η is deformed into the germ of f just described.

It is possible to define mappings between certain unfoldings and hence to construct a category of unfoldings of a singularity. The objects of the category are the unfoldings. To motivate the definition of morphism observe first that $\mathbf{R}^{n+r}$ is fibred by the projection $\pi_r : \mathbf{R}^{n+r} \to \mathbf{R}^r$. Any mapping between unfoldings should respect the fibration because on the fibres there are the germs 'near η' defined in the preceding paragraph. Amongst the family of all such germs on fibres $\pi_r^{-1}(u) = \mathbf{R}^n \times \{u\}$ near $\pi_r^{-1}(0) = \mathbf{R}^n \times \{0\} = \mathbf{R}^n$, one finds at least some of the singularities which are concealed in η, together with their unfoldings. This structure should be preserved.

The morphism is permitted to transform arbitrarily the parameter space $\mathbf{R}^r$ and the fibres $\mathbf{R}^r \times \{u_0\}$ for each u_0 other than the origin. In the range space $\mathbf{R}$ one should also allow arbitrary transformations, but the present study will be restricted to the simpler case where only translations are allowed. The definition is as follows:

14.2. **Definition.** Let (r, f) and (s, g) be unfoldings of η. A morphism

$$(\phi, \alpha) : (r, f) \to (s, g)$$

consists of

(I) A germ $\phi \in \mathcal{E}(n+r, n+s)$ with $\phi|\mathbf{R}^n \times \{0\} = \mathrm{id}$,

(II) A germ $\Phi \in \mathcal{E}(r, s)$, such that $\pi_s \phi = \Phi \pi_r$,

(III) A germ $\alpha \in \mathfrak{m}(r)$, such that

$$f = g \circ \phi + \alpha \circ \pi_r.$$

Taken together, the conditions (I) and (II) say that ϕ is represented by a 'fibrewise' map $\mathbf{R}^{n+r} \to \mathbf{R}^{n+s}$:

$$\phi(x, u) = (\phi_1(x, u), \Phi(u))$$

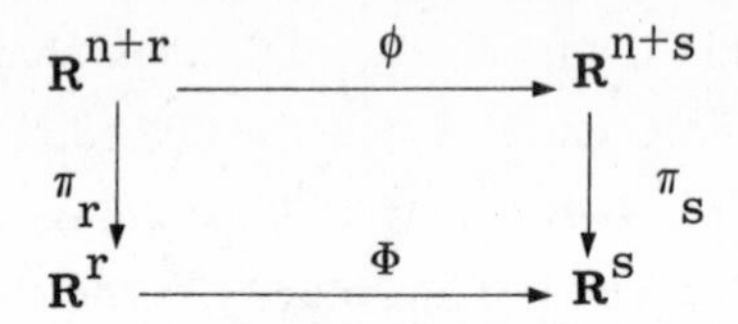

Given a representative for α, one may assign a translation a_u to each $u \in \mathbf{R}^r$ near the origin. The translation $a_u : \mathbf{R} \to \mathbf{R}$ is given by $a_u(t) = t + \alpha(u)$. Condition (III) in the definition states that the germs f and g are related by

$$f(x, u) = a_u \circ g \circ \phi(x, u).$$

This is an obvious definition, however it would be more plausible to allow an arbitrary family of transformations a_u instead of just translations (or, alternatively, not to transform the range at all). The definition given here follows Mather and is relatively simple. At least one may suitably adjust the origin for each u using translations. The general case is dealt with by Wassermann.

Morphisms are composed in the obvious way,

$$(\phi, \alpha)(\psi, \beta) = (\phi \circ \psi, \beta + \alpha \circ \Psi).$$

At a point $u \in \mathbf{R}^r$ of the second parameter space $\beta(u) + \alpha\Psi(u)$ describes the composition of translations, namely $a_{\Psi(u)} \circ b_u$.

A morphism (ϕ, α) is obviously invertible (an isomorphism) exactly when ϕ is invertible, in particular,

$$(\mathrm{id}, \alpha) : (r, f) \cong (r, f + \alpha).$$

The function α allows the germ on the fibre over u to be displaced by an additive constant $\alpha(u)$.

14.3. Addition of unfoldings

$$(r, f) + (s, g) = (r+s, f+g - \eta)$$

where the last term is given by

$$(f + g - \eta)(x, u, v) = f(x, u) + g(x, v) - \eta(x).$$

14.4. The constant unfolding (r, η) is defined by

$$\eta(x, u) = \eta(x).$$

One finds

$$(r, f) + (s, \eta) = (r+s, f).$$

The formula in (III) in the definition of a morphism shows that the unfolding (r, f) is determined by the morphism (ϕ, α) and the unfolding (s, g). Hence we make the following

14.5. **Definition.** Let (s, g) be an unfolding of η and suppose the germs $\phi \in \mathcal{E}(n+r, n+s)$ and $\Phi \in \mathcal{E}(r, s)$ satisfy (I) and (II) in 14.2. Then if $\alpha \in \mathfrak{m}(r)$, the unfolding (r, f) given by equation (III) is called the unfolding of η induced by (ϕ, α) from (s, g).

An unfolding (r, f) of η is called versal if any unfolding of η is induced from (r, f) by a suitable morphism.

14.6. **Example of an unfolding.** Let $\eta \in \mathfrak{m}(n)$ be a singularity, and let $b_1, \ldots, b_r \in \mathfrak{m}(n)$, then

$$f(x, u) = \eta(x) + b_1(x)u_1 + \ldots + b_r(x)u_r$$

is an unfolding of η.

This unfolding is the sum of the one-parameter unfoldings $f(x, u_i) = \eta(x) + b_i(x)u_i$.

14.7. **Definition.** For a singularity η, define the codimension of η by $\operatorname{codim} \eta = \dim_{\mathbf{R}}(\mathfrak{m}(n)/\langle \partial\eta/\partial x_i \rangle_{\mathcal{E}(n)})$.

A versal unfolding (r, f) with minimal r is called universal.

The significance of universal unfoldings and of the codimension becomes clear in the main theorem about unfoldings:

14.8. **Theorem (Mather).** A singularity $\eta \in \mathfrak{m}(n)$ has a versal unfolding if and only if $\operatorname{codim} \eta$ is finite.

Two r-parameter versal unfoldings are isomorphic.

Every versal unfolding is isomorphic to (r, f) + constant, where (r, f) <u>is universal.</u>

<u>If</u> $\{b_1, \ldots, b_r\} \subset m(n)$ <u>is a system of representatives for a basis of</u> $m(n)/\langle \partial\eta/\partial x_i \rangle_{\mathcal{E}(n)}$, <u>then the unfolding</u> f <u>of</u> η <u>defined by</u>

$$f(x, u) = \eta(x) + b_1(x)u_1 + \ldots + b_r(x)u_r$$

<u>is universal.</u>

(The proof takes up all of chapter 16.)

14.9. Example. From now on, abbreviate $\langle \partial\eta/\partial x_i \rangle_{\mathcal{E}(n)}$ simply to $\langle \partial\eta \rangle$. Let $n = 1$ and $\eta(x) = x^N$, then $\langle \partial\eta \rangle = \langle x^{N-1} \rangle$ and $m/\langle \partial\eta \rangle$ has the basis $x, x^2, \ldots, x^{N-2}$. Therefore

$$f(x, u) = x^N + u_{N-2}x^{N-2} + u_{N-3}x^{N-3} + \ldots + u_1 x$$

is the universal unfolding of x^N.

More generally, let

$$\eta(x) = x_1^N \pm x_2^2 \pm x_3^2 + \ldots \pm x_n^2,$$

then $\langle \partial\eta \rangle = \langle x_1^{N-1}, x_2, \ldots, x_n \rangle$, so that the universal unfolding is

$$f(x, u) = \eta(x) + u_{N-2}x_1^{N-2} + \ldots + u_1 x_1.$$

This last comment can be formulated in general:

14.10. Remark. If $\eta(x_1, \ldots, x_k)$ has a universal unfolding $\eta + f(x_1, \ldots, x_k, u)$, and if $q(x_{k+1}, \ldots, x_n)$ is a non-degenerate quadratic form in further variables, then $\eta + q$ has $\eta + q + f(x_1, \ldots, x_k, u)$ as its universal unfolding. This is because q may be put in the form $\pm x_{k+1}^2 \pm \ldots \pm x_n^2$ after a suitable linear transformation, and therefore $\langle \partial(\eta + q) \rangle = \langle \partial\eta, x_{k+1}, \ldots, x_n \rangle$. This means that $m(n)/\langle \partial(\eta + q) \rangle$ has the 'same basis' as $m(k)/\langle \partial\eta \rangle$.

Thus, when one computes a universal unfolding, it is convenient to first transform the singularity in such a way that as many variables as possible are separated out into a quadratic form.

14.11. Definition. The corank of a singularity $\eta \in \mathfrak{m}(n)$ is the corank of the Hessian $(\partial^2\eta/\partial x_i \partial x_j(0))$, i.e. the quadratic form given by the two jet.

14.12. Splitting lemma. If $\eta \in \mathfrak{m}(n)$ is a singularity with corank $n - r$ then η is right-equivalent to a germ of the form

$$q(x_1, \ldots, x_r) + \zeta(x_{r+1}, \ldots, x_n)$$

where $j^2\zeta = 0$, and q is a non-degenerate quadratic form.

Proof. In any case, after a linear transformation, the 2-jet of η may be given the form $q(x_1, \ldots, x_r) = \pm x_1^2 \pm \ldots \pm x_r^2$. Define $\theta = \eta|\mathbf{R}^r$, so that θ has the 2-jet q. It follows that θ is 2-determined (11.3) and hence right-equivalent to q. Therefore one may assume $\eta|\mathbf{R}^r = q$. Obviously q has the universal unfolding $(0, q)$, since $\langle\partial q\rangle = \langle x_1, \ldots, x_r\rangle$. Hence η is a versal unfolding of q, since it contains the universal unfolding. Because versal unfoldings of fixed dimension are isomorphic (main theorem), $(n - r, q)$ is isomorphic to $(n - r, \eta)$, that is, there exist $-\zeta \in \mathfrak{m}(n - r)$ and an invertible germ ϕ so that

$$q(x_1, \ldots, x_r) = \eta\phi(x_1, \ldots, x_n) - \zeta(x_{r+1}, \ldots, x_n).$$

This completes the proof of the splitting lemma. ✓

Recall that the codimension of the germ η is $\dim_{\mathbf{R}} \mathfrak{m}(n)/\langle\partial\eta\rangle$. The germs of greatest interest are those whose universal unfoldings contain at most four parameters, that is, those η with codim $\eta \leq 4$.

14.13. Remark. If corank $\eta = r$, then codim $\eta \geq \binom{r+1}{2}$. In particular corank $\eta \geq 3 \Rightarrow$ codim $\eta \geq 6$.

Proof. If corank $\eta = r$, then, using the splitting lemma, there are suitable coordinates for which

$$\eta(x) = \zeta(x_1, x_2, x_3, \ldots, x_r) + q(x_{r+1}, \ldots, x_n)$$

with $j^2\zeta = 0$.

Consider the quotient of the vector space $\mathfrak{m}(n)/\langle\partial\eta\rangle$ obtained by restricting to the first r coordinates and taking 2-jets, i.e. consider $j^2\,\mathfrak{m}(r)/j^2\langle\partial\zeta\rangle$. This, then, has dimension smaller than codim η. Now $\dim j^2\,\mathfrak{m}(r) = r + \binom{r+1}{2}$ and $\dim j^2\langle\partial\zeta\rangle \leq r$: the $\partial\zeta/\partial x_i$ themselves generate the real vector space $j^2\langle\partial\zeta\rangle$ because each $\partial\zeta/\partial x_i$ belongs to $\mathfrak{m}(r)^2$. Hence

$$\text{codim}\,\eta \geq \dim j^2\,\mathfrak{m}(r)/j^2\langle\partial\zeta\rangle \geq \binom{r+1}{2}. \quad \checkmark$$

Thus if one is interested in universal unfoldings with at most 5 parameters, one may restrict attention to singularities $\eta(x, y)$ in two variables which have codimension ≤ 5 (the non-degenerate quadratic form q does not affect the unfolding). These singularities are automatically finitely determined, in fact 6-determined. In suitable coordinates they can be written as polynomials in 2 variables with degree at most 6.

In the study of finitely determined germs the dimension of $\mathfrak{m}(n)/\mathfrak{m}(n)\langle\partial\eta\rangle$ has been used. In chapter 11 we observed that

$$\mathfrak{m}(n)\langle\partial\eta\rangle + \langle\partial\eta\rangle_{\mathbf{R}} = \langle\partial\eta\rangle_{\mathcal{E}(n)}$$

and so it follows that

14.14. $\dim(\mathfrak{m}(n)/\mathfrak{m}(n)\langle\partial\eta\rangle) \leq \dim(\mathfrak{m}(n)/\langle\partial\eta\rangle) + n$ with equality holding when the $\partial\eta/\partial x_i$ modulo $\mathfrak{m}(n)/\langle\partial\eta\rangle$ are linearly independent. In particular, η is finitely determined, i.e. $\dim(\mathfrak{m}(n)/\mathfrak{m}(n)\langle\partial\eta\rangle) < \infty$, if and only if codim $\eta < \infty$ (see 11.10).

14.15. Lemma. If the codimension of η is finite (η finitely determined) then $\dim(\mathfrak{m}(n)/\mathfrak{m}(n)\langle\partial\eta\rangle) = \text{codim}\,\eta + n$.

As we know from lemma 11.8, the left-hand side describes the codimension of the orbit of η under right-transformation in the space $\mathfrak{m}(n)$. In chapter 16 we shall see that an r-parameter family of functions defines a germ $(\mathbf{R}^{n+r}, 0) \to \mathfrak{m}(n)$. If a 'general' r-parameter family contains η, then this germ will be transversal to the orbit of η. The lemma shows that the codimension of η should not exceed r if η is to have a generic unfolding with r parameters.

Proof of the lemma. It has to be proved that the $\partial\eta/\partial x_i$ are linearly independent modulo $m(n)\langle\partial\eta\rangle$. If this were not so, there would be $a_i \in \mathbf{R}$ and germs $u_i \in m(n)$ such that

$$\sum a_i \frac{\partial\eta}{\partial x_i} = \sum u_i \frac{\partial\eta}{\partial x_i},$$

that is,

$$\sum (a_i - u_i) \frac{\partial\eta}{\partial x_i} = 0,$$

$$a_i - u_i(0) \neq 0 \text{ for some } i.$$

Integrate the vector field $\sum(a_i - u_i)\partial/\partial x_i$ and choose coordinates so that the integral curves are given by $x_i = c_i$, for $i > 1$ and constants c_i. That is, so that the vector field is equal to $\partial/\partial x_1$. It follows that $\partial\eta/\partial x_1 \equiv 0$. This makes η independent of x_1 and hence $x_1, x_1^2, x_1^3, \ldots$ are independent in $m(n)/\langle\partial\eta\rangle$. But then codim η would be ∞, contradiction. ✓

Two problems are left outstanding, one is the complete classification of the singularities of codimension ≤ 4, with their universal unfoldings (chapter 15). The second is the proof of the main theorem on unfoldings (chapter 16).

15 · The seven elementary catastrophes

Literature: as for chapter 14.

If η is a singularity with codimension ≤ 4, we know

$$\text{codim } \eta \leq 4 \iff \dim(\mathfrak{m}(n)/\langle\partial\eta\rangle) \leq 4$$

$$\Rightarrow \quad \mathfrak{m}(n)^5 \subset \langle \partial\eta/\partial x_i \rangle \Rightarrow \mathfrak{m}(n)^6 \subset \mathfrak{m}(n)\langle\partial\eta\rangle$$

$\Rightarrow \quad \eta$ is 6-determined.

Hence in some coordinate frame η is a polynomial of degree ≤ 6 in two variables plus a non-degenerate quadratic form in other variables (see 14.13). We must now transform such polynomials into normal form using further coordinate changes. The result is as follows:

15.1. Theorem: the rule of seven (Thom). Up to the addition of a non-degenerate quadratic form in other variables and up to multiplication by ± 1, a singularity of codimension ≤ 4 and ≥ 1 is right-equivalent to one of the following:

Codim	η	Universal Unfolding	Name
1	x^3	x^3+ux	Fold
2	x^4	x^4-ux^2+vx	Cusp (Riemann-Hugoniot)
3	x^5	$x^5+ux^3+vx^2+wx$	Swallowtail or Dovetail
3	x^3+y^3	$x^3+y^3+wxy-ux-vy$	Hyperbolic Umbilic
3	x^3-xy^2	$x^3-xy^2+w(x^2+y^2)-ux-vy$	Elliptic Umbilic
4	x^6	$x^6+tx^4+ux^3+vx^2+wx$	Butterfly
4	x^2y+y^4	$x^2y+y^4+wx^2+ty^2-ux-vy$	Parabolic Umbilic

Proof. It follows directly from the main theorem that the universal unfoldings have these forms. We must show that the list contains all possible germs.

1. Corank $\eta = 1$.

Here η is right-equivalent to $\pm x^n$ (up to a quadratic form). Hence if the codimension is ≤ 4, only x^3, x^4, x^5 and x^6 are possible (up to multiplication by ± 1).

2. Corank $\eta = 2$.

This implies codim $\eta \geq 3$ (14.13), so codim η is 3 or 4.

Let $P(x, y) = j^3(\eta)$.

This is obviously a homogeneous polynomial of degree 3 and hence P splits over $\mathbf{C}$ into three linear factors:

$$P(x, y) = (a_1 x + b_1 y)(a_2 x + b_2 y)(a_3 x + b_3 y).$$

There are the following four possibilities which we shall discuss separately.

(A) The three vectors $(a_i, b_i) \in \mathbf{C}^2$ are pairwise linearly independent over $\mathbf{C}$.

(B) Without loss of generality, the first two vectors are linear independent and the third is a multiple of the second. Here $P(x, y) = (a_1 x + b_1 y)(a_2 x + b_2 y)^2$ with (a_1, b_1), (a_2, b_2) linearly independent. Because the factorisation is unique up to constants and P is real, the factors and hence the (a_i, b_i) may be chosen real (consider the conjugate factorisation).

(C) All the (a_i, b_i) are dependent, but all $\neq 0$. Then

$$P(x, y) = (ax + by)^3, \ (a, b) \in \mathbf{R}^2.$$

(D) $P(x, y) = 0$.

Case (A).

(α) Suppose all the (a_i, b_i) are real: choose $(a_1 x + b_1 y)$, $(a_2 x + b_2 y)$ as new coordinates. Writing $\sim$ for right-equivalence, we have

$$P(x, y) \sim xy(ax + by) \text{ with } a, b \neq 0.$$

Now

$$\begin{aligned} xy(ax + by) &\sim (ab)^{-1}xy(x + y) && \text{using } (x, y) \mapsto (ax, by) \\ &\sim xy(x + y) && \text{using } (x, y) \mapsto (ab)^{-1/3}(x, y) \\ &\sim x(x^2 - y^2) && \text{using } (x, y) \mapsto 2^{-2/3}(x + y, x - y) \\ &= x^3 - xy^2. \end{aligned}$$

This polynomial is 3-determined so $\eta \sim x^3 - xy^2$ (elliptic umbilic).

(β) Suppose two (a_i, b_i) are complex conjugates. Then

$$P(x, y) = (a_1x + b_1y)(a_2x + b_2y)(\bar{a}_2x + \bar{b}_2y).$$

The product of the last two factors is a positive definite quadratic form in x, y. After a change of coordinates this may be written $x^2 + y^2$ and therefore $P(x, y) \sim (ax + by)(x^2 + y^2)$. By rotating coordinates, $(ax + by)$ may be transformed into cx, $c \neq 0$. Then

$$P \sim cx(x^2 + y^2) \sim x(x^2 + y^2) \sim x^3 + xy^2 \sim x^3 + y^3.$$

The last equivalence follows because

$$(x + y)^3 + (x - y)^3 = 2x^3 + 6xy^2 \sim x^3 + xy^2.$$

$x^3 + y^3$ is also 3-determined. Hence $\eta \sim x^3 + y^3$ (hyperbolic umbilic). Case (A). √

Case (B)

$$P(x, y) = (a_1x + b_1y)(a_2x + b_2y)^2 \sim x^2y.$$

Notice that x^2y is not finitely determined, since $\partial/\partial x(x^2y) = 2xy$, $\partial/\partial y(x^2y) = x^2$ and the ideal $\langle xy, x^2\rangle$ does not contain any powers of y. However η is finitely determined and so it must have a jet which is not equivalent to x^2y. Suppose k is the largest number for which $j^k\eta \sim x^2y$. Without loss, $j^k\eta = x^2y$ and $j^{k+1}\eta = x^2y + h(x, y)$, where h is a homogeneous polynomial of degree $k + 1$, $k \geq 3$. We transform η with a diffeomorphism of the form $\Phi : (x, y) \to (x + \phi, y + \psi)$ where

ϕ, ψ are homogeneous of degree $k - 1 \geq 2$. The Jacobian of Φ at the origin is the identity. We obtain

$$j^{k+1}\eta \circ \Phi = x^2y + x^2\psi + 2xy\phi + h(x, y).$$

By suitable choice of ϕ, ψ, one may eliminate all the terms in h which are divisible by xy or x^2. This gives

$$j^{k+1}\eta \circ \Phi = x^2y + ay^{k+1}, \ a \neq 0.$$

One easily checks that this is (k+1)-determined and hence $\eta \sim x^2y + ay^{k+1} \sim x^2y \pm y^{k+1}$. If $k \geq 4$ then codim $\eta \geq 5$. Hence $k = 3$ and $x^2y + y^4 \sim x^2y - y^4$ (multiply by -1 and replace y by -y).
Case (B). ✓

Case (C)

$P = (ax + by)^3 \sim x^3$ so that, without loss, $j^3\eta = x^3$. Then $j^4\eta = x^3 + h$ where h has degree 4. One checks that

$$\dim j^3\mathfrak{m}(2) = 9$$
$$\dim j^3\langle\partial\eta\rangle = \dim j^3\langle x^2 + h_1, h_2\rangle \leq 4$$

(degree h_1, degree $h_2 \geq 3$).

Hence $\dim j^3\mathfrak{m}(2)/\langle\partial\eta\rangle \geq 5 > 4$.
This case cannot satisfy codim $\eta \leq 4$. ✓

Case (D)

$P = 0$ implies $\eta \in \mathfrak{m}(2)^4$. Hence $\langle\partial\eta\rangle \subset \mathfrak{m}(2)^3$ and $\dim(\mathfrak{m}(2)/\mathfrak{m}(2)^3) = 5$. This case is also excluded. ✓

The proof of the seven-catastrophe theorem is now complete. ✓

The significance of this result is as follows:

One imagines some sort of chemical system e.g. described by n variables, that is, by a point $x \in \mathbf{R}^n$. The system is subject to a dynamic flow, which is described by a potential function $\tilde{V} : X \to \mathbf{R}$ when the 'external conditions' remain fixed. Now suppose that the external conditions vary - that they vary according to place and time. The alteration in the conditions is accompanied by a change in the potential function. For each point u in an open subset $U \subset \mathbf{R}^4$ (space-time) one has a potential function $V_u : X \to \mathbf{R}$. Hence, there is a (differentiable) map

$$V : X \times U \to \mathbf{R},$$

a family of potential functions on X, parametrised by U. When the external conditions are fixed, at a fixed point $u \in U$, the system will stay in a minimum of the corresponding potential function V_u. Usually this will be a non-degenerate singular point.

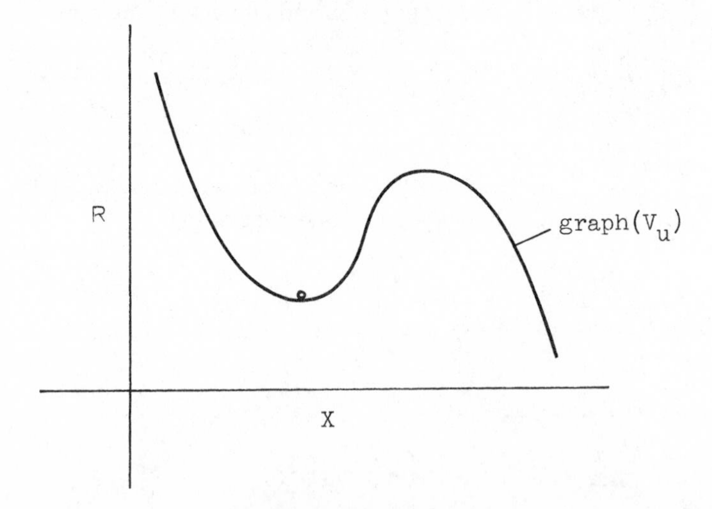

Of course there are potential functions with degenerate singular points, but these are 'improbable'. One can prove fairly easily that the Morse-functions form an open, dense subset in the set of all functions. Morse-functions are characterised as follows:

(i) At a singular point x_0 of V, the quadratic form of second derivatives

$$\left(\frac{\partial^2 V}{\partial x_i \partial x_j}(x_0)\right)$$

is non-degenerate. In particular, this implies that V is 2-determined at x_0 and will have the form

$$V(x) = V(0) + \sum \pm x_i^2$$

with respect to suitable coordinates. Further, the singularities are obviously isolated. The second requirement for a Morse-function is

(ii) if $x \neq y$ are singular points, then $V(x) \neq V(y)$.

In general, then, a potential function V_u is a Morse-function. However as u varies, for example as it moves in the 1-dimensional time-subspace, one may ask which kinds of singularities can occur generically in this family of functions V_u.

The local change in V_u around a point $u_0 \in U$ and a singular point $x_0 \in \mathbf{R}^n$ corresponds exactly to an unfolding of the germ V_{u_0} around x_0. The versal unfoldings give a description of all possible unfoldings. Furthermore, if one defines a concept of 'stable' unfolding in a natural (though rather complicated) manner, it turns out that the seven catastrophes are also the only possible stable unfoldings with codimension ≤ 4 (see Wassermann).

With a view to applications it is interesting to describe the geometric appearance of the seven universal unfoldings of codimension ≤ 4 more precisely. In particular one wants to see those points in the control space - the space of unfolding parameters U - which are most significant for the catastrophe. These are the points where V_u has a singularity of order higher than two. In other words, interest centres on those points where a local minimum (or maximum) disappears.

For the cusp one obtains the picture on the next page (the potential function is drawn for 5 points in U).

If one places the x-coordinate perpendicular to the (u, v)-coordinate system, then the local extrema lie on a surface $\{(x, u, v) | 4x^3 - 2ux + v = 0\}$. The projection onto the (u, v)-plane shows the familiar cusp as its set of singular values. Any state, whose (u, v)-parameter moves across the upper branch of the cusp from below, suddenly jumps out of the minimum belonging to the upper surface into that of the lower surface. A reversed process occurs on crossing the lower branch.

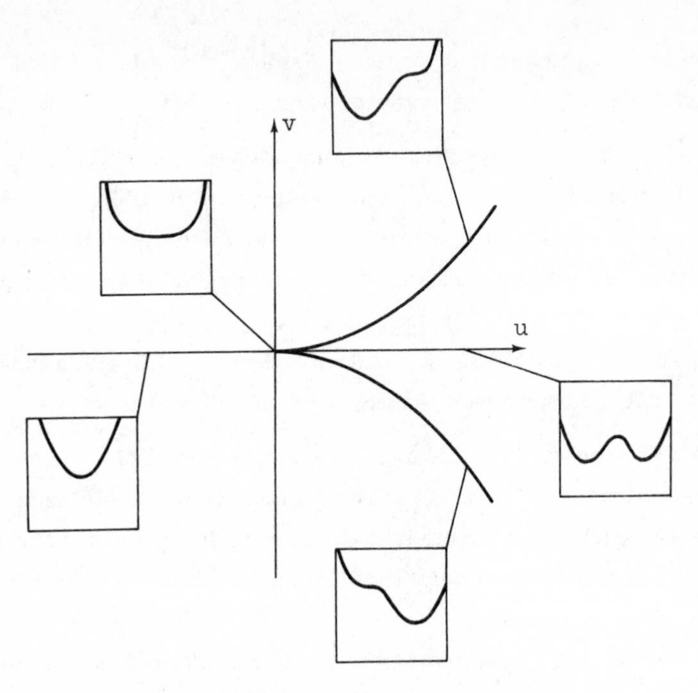

Note: The u in the text is (u, v) in the example.

This example is discussed in more detail together with the other elementary catastrophes in chapter 17.

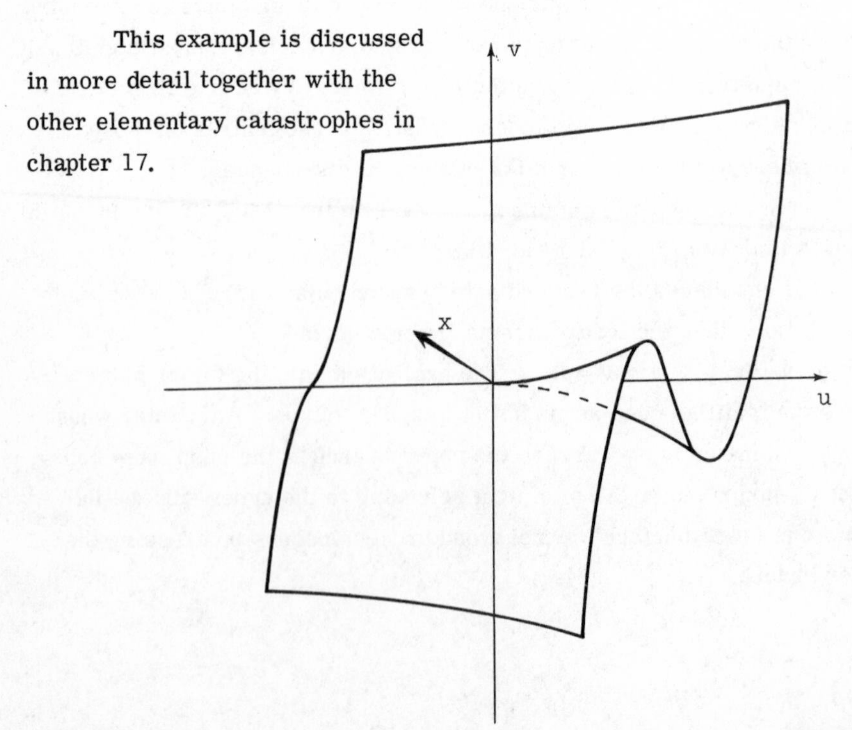

16 · Proof of the main theorem on universal unfoldings

Literature: as for chapter 14.

Throughout this chapter η will be a singularity. The versal unfoldings of η will be characterised by a transversality condition which we must first explain.

Let $\eta \in m(n)^2$ be a germ and $(r, \tilde{f})$ an r-parameter unfolding of η. Suppose f is a representative of $\tilde{f}$. If $J_0^k(n, 1)$ is the space of k-jets whose zero-th term vanishes, we may define a germ

$$j_1^k f : (\mathbf{R}^{n+r}, 0) \to J_0^k(n, 1)$$

in the following way:

A representative of $j_1^k f$ is the map $(x, u) \mapsto$ k-jet of $(y \mapsto f(x + y, u) - f(x, u))$.

Thus, $j_1^k f$ is a generalised partial derivative: the partial Taylor expansion at the point (x, u) with respect to the first variables.

16.1. **Definition.** f is called k-transversal, if $j_1^k f$ is transversal at the origin to the orbit $\hat{\eta}\hat{\mathcal{B}}_k(n)$ of $\hat{\eta}$ (= k-jet of η) under right transformation.

Obviously $j_1^k f(0) = j^k \eta(0) = \hat{\eta} \in \hat{\eta}\hat{\mathcal{B}}_k(n)$.

We may add the following characterisation of versal unfoldings to the main theorem.

16.2. **Theorem** (versal = k-transversal). If η is k-determined, then an unfolding of η is versal if and only if it is k-transversal.

The hardest part of the chapter will be the proof of the next lemma.

16.3. **Main Lemma.** If η is k-determined and (r, f), (r, g) are k-transversal unfoldings, then $(r, f) \cong (r, g)$.

We shall first interpret the concept of k-transversality using explicit formulae. In this way we shall be able to deduce the main theorem (14.8) and the theorem above (16.2) from the main lemma. The proof of the main lemma will be left till last.

16.4. Lemma. f _is_ k-_transversal if and only if_

$$\mathfrak{m}(n) = \langle \partial\eta/\partial x_i \rangle + V_f + \mathfrak{m}(n)^{k+1}$$

where $V_f = \langle \partial f/\partial u_j | \mathbf{R}^n \times \{0\} - \partial f/\partial u_j(0) \rangle_{\mathbf{R}}$ _is the real vector space spanned by the given elements._

Proof. We know (11.8) that the tangent space to $\hat{\eta}\hat{\mathcal{B}}_k$ is

$$\mathfrak{m}(n)\langle \partial\eta/\partial x_i \rangle + \mathfrak{m}(n)^{k+1}.$$

We must compute the image of $Dj_1^k f(0)$. The tangent space $T_0(\mathbf{R}^n \times \mathbf{R}^r)$ is spanned by $\frac{\partial}{\partial x_i}, \frac{\partial}{\partial u_j}$ and so the image is spanned by

$$\frac{\partial}{\partial x_i} j_1^k f(0) = j_1^k \frac{\partial}{\partial x_i} f(0) \text{ and } j_1^k \frac{\partial}{\partial u_j} f(0).$$

Now

$$j_1^k \frac{\partial}{\partial x_i} f(0) = j_1^k \frac{\partial}{\partial x_i} \eta(0).$$

The lemma follows from the equalities

$$\mathfrak{m}(n)\langle \partial\eta \rangle + \mathfrak{m}(n)^{k+1} + \langle j^k \frac{\partial \eta}{\partial x_i}(0) \rangle_{\mathbf{R}} = \langle \partial\eta \rangle + \mathfrak{m}(n)^{k+1}$$

and

$$\langle j_1^k \frac{\partial}{\partial u_j} f(0) \rangle_{\mathbf{R}} + \mathfrak{m}(n)^{k+1} = \langle \frac{\partial f}{\partial u_j} | \mathbf{R}^n \times \{0\} - \frac{\partial f}{\partial u_j}(0) \rangle_{\mathbf{R}} + \mathfrak{m}(n)^{k+1}. \checkmark$$

16.5. Corollary. _If_ $b_1, \ldots, b_r$ _is a basis of_ $\mathfrak{m}(n)/(\langle \partial\eta \rangle + \mathfrak{m}(n)^{k+1})$, _then_ $\eta + \sum u_j b_j$ _is_ k-_transversal._ ✓

16.6. Corollary. _If_ (r, f) _is a versal unfolding of_ η, _then_ f _is_ k-_transversal for every_ k.

Proof. Choose a k-transversal unfolding (s, g), this is easily found using the first corollary (16.5). There must be a morphism

$$(\phi, \alpha) : (s, g) \to (r, f).$$

Thus $g = f \circ \phi + \alpha$, and since α is not dependent on x it follows that $V_g = V_{f\phi}$. Now,

$$V_{f\phi} \subset \langle \partial\eta \rangle + V_f$$

because

$$\frac{\partial f \circ \phi}{\partial u_j} = \sum_{i=1}^{n} \frac{\partial f}{\partial x_i} \frac{\partial \phi_i}{\partial u_j} + \sum_{\nu=1}^{r} \frac{\partial f}{\partial v_\nu} \frac{\partial \phi_\nu}{\partial u_j}$$

and so restricted to $\mathbf{R}^n \times \{0\}$:

$$\frac{\partial f \circ \phi}{\partial u_j} = \sum_{i=1}^{n} \frac{\partial \eta}{\partial x_i} \cdot \frac{\partial \phi_i}{\partial u_j} + \sum_{\nu=1}^{r} \frac{\partial f}{\partial v_\nu} \cdot a_{\nu j}$$

with $a_{\nu j} = \frac{\partial \phi_\nu}{\partial u_j}(0)$.

It follows from lemma 16.4 that f is k-transversal. √

Versal unfoldings, then, are k-transversal. The converse is even easier - using the main lemma.

Proof of 16.2 (versal = k-transversal). Let η be k-determined, (r, f) k-transversal and (s, g) an arbitrary unfolding. We have to find a morphism $(s, g) \to (r, f)$. This is it:

$(s, g) \xrightarrow{\text{obvious}} (s, g) + (r, f)$, k-transversal because f is,

$\xrightarrow[\text{lemma}]{\text{main}}$ const. $+ (r, f)$, k-transversal,

$\xrightarrow{\text{obvious}} (r, f)$. √

16.7. **Corollary** (compare lemma 14.15). <u>If</u> (r, f) <u>is a versal unfolding of</u> η, <u>then</u> codim $\eta \leq r$.

Proof. (r, f) is k-transversal and so $\mathfrak{m}(n) = \langle \partial\eta \rangle + V_f + \mathfrak{m}(n)^{k+1}$. Therefore $\dim(\mathfrak{m}(n)/(\langle \partial\eta \rangle + \mathfrak{m}(n)^{k+1})) \leq \dim V_f \leq r$. This is satisfied for all k and hence, by Nakayama's lemma,

$$\mathfrak{m}(n)^k \subset \langle \partial\eta \rangle + \mathfrak{m}(n)^{k+1} \quad \text{for } k > r.$$

Applying Nakayama's lemma once more gives $\mathfrak{m}(n)^k \subset \langle \partial\eta \rangle$ and so $\dim \mathfrak{m}(n)/\langle \partial\eta \rangle \leq r$. √

Proof of the main theorem (14. 8). A complete proof of the main theorem on unfoldings can now be given (still assuming the main lemma).

If η is k-determined, then two r-parameter versal unfoldings of η are k-transversal. Hence the two unfoldings are isomorphic. If (r, f) is a versal unfolding of smallest dimension then (r, f) and therefore (r, f) + constant are both k-transversal. We deduce that all k-transversal unfoldings can be obtained from (r, f). Corollary 16. 7 shows that the smallest possible number of parameters in a versal unfolding is codim η. If the codimension is finite (equal to r say) then η is finitely determined (c. f. 11. 4). Corollary 16. 5 gives an r-parameter, k-transversal, hence universal, unfolding of η with the required form. √

Now we come to the hardest part of our task.

Proof of the main lemma (16. 3). Let η be a k-determined singularity, (r, f), (r, g) k-transversal unfoldings of η. We must find an isomorphism $(r, f) \cong (r, g)$. We know that (r, f) is k-transversal if

$$\langle \partial\eta/\partial x_i \rangle + V_f + \mathfrak{m}(n)^{k+1} = \mathfrak{m}(n)$$

where V_f is generated over $\mathbf{R}$ by the terms

$$\left(\frac{\partial f}{\partial u_i}\,\Big|\,\mathbf{R}^n \times \{0\} - \frac{\partial f}{\partial u_i}(0)\right).$$

Problem. We seek a homotopy F_t of transversal unfoldings starting at $F_0 = f$ and ending at $F_1 = g$ (afterwards we show that F_t is locally constant as a map from $[0, 1]$ into the isomorphism types of unfoldings).

Solution of this problem. The r-parameter unfoldings of η are the germs in $\eta + \mathfrak{m}(r).\mathscr{E}(n + r) \subset \mathfrak{m}(n + r)$, where $\mathfrak{m}(r)$ is generated by $u_1, \ldots, u_r$. Thus, they may be written $\eta + \delta$ where $\delta \in \mathfrak{m}(r).\mathscr{E}(n + r)$ and it is clear that $V_{\eta+\delta} = V_\delta$. In the space

$J^k_0(n)$ of k-jets with vanishing constant term, there is the subspace $\langle \partial\eta/\partial x_i \rangle / \mathfrak{m}(n)^{k+1}$, and we are interested in those δ for which V_δ is transversal to this subspace.

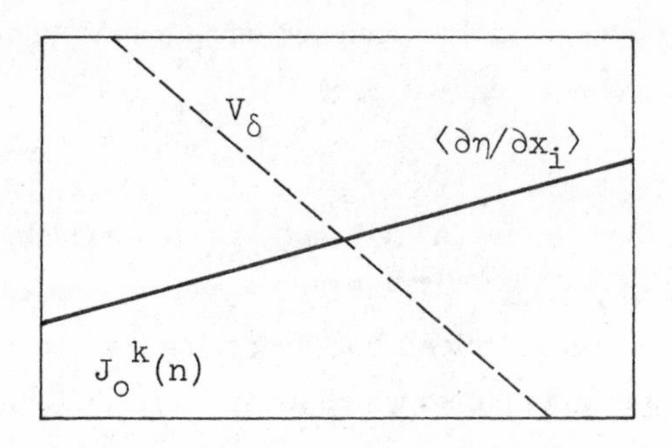

We introduce the map

$$\mathfrak{m}(r)\mathscr{E}(n+r) \to \mathrm{Hom}(\mathbf{R}^r, J^k_0(n))$$

$$\delta \mapsto (e_i \mapsto j^k(\frac{\partial\delta}{\partial u_i}|\mathbf{R}^n \times \{0\} - \frac{\partial\delta}{\partial u_i}(0)))$$

where the e_i are basis elements of $\mathbf{R}^r$. This map is obviously surjective since suitable polynomials can be chosen for δ.

Consider the exceptional subset A in Hom consisting of those homomorphisms for which the image of $\mathbf{R}^r$ is not transversal to $\langle \partial\eta/\partial x_i \rangle \, \mathfrak{m}(n)^{k+1}$. It is clear that A is algebraic.

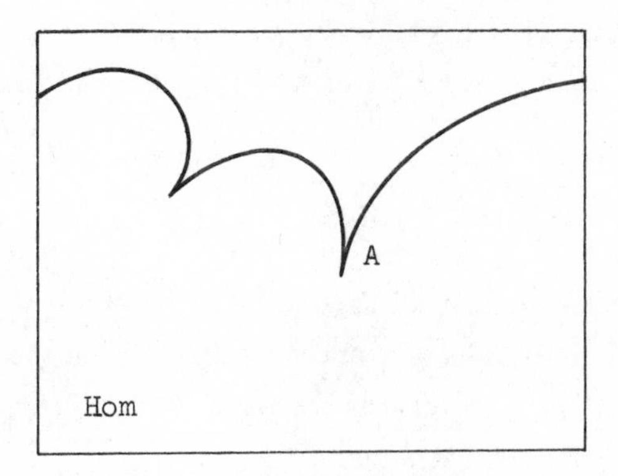

If $\operatorname{codim} \eta = s = \operatorname{codim}\langle \partial\eta/\partial x_i \rangle$ in J_0^k, then we know that $s \leq r$ by corollary 16.7, which was actually proved for k-transversal unfoldings.

Case 1

Let $r > s$. One easily convinces oneself that $\operatorname{codim} A > 1$ and then Hom - A is connected, (use 9.3).

Case 2

If $r = s$, then Hom - A falls into two components according to the orientation of the image of $\mathbf{R}^r$ with respect to $\langle \partial\eta/\partial x_i \rangle \mathfrak{m}(n)^{k+1}$. However if $\phi \in \mathcal{B}(r)$ is an orientation-reversing transformation, then $\eta + \delta$ and $(\eta + \delta)\phi$ give points which are in different components of Hom - A. Thus we may assume that f and g are mapped into the same component of Hom - A.

It follows that the images of f and g can be joined by a polygonal path in Hom - A. It is obvious that a linear path in Hom - A can be lifted to a linear path in $\mathfrak{m}(r)\mathcal{E}(n + r)/\mathfrak{m}(n + r)^{k+1}$ (this maps surjectively onto Hom). This path lifts to a linear path in $\eta + \mathfrak{m}(r)\mathcal{E}(n+r)$. Hence f and g can be joined piecewise linearly by k-transversal unfoldings and without loss of generality, we can assume that

$$F_t = (1 - t)f + tg \text{ is k-transversal for } 0 \leq t \leq 1.$$

We have now to show that the isomorphism type of F_t as an unfolding is locally constant.

Without loss, $f | \{0\} \times \mathbf{R}^r = g | \{0\} \times \mathbf{R}^r = 0$ since putting $\alpha_t(u) = (1 - t)f(0, u) + tg(0, u)$ gives an isomorphism (Id, α_t) between F_t and

$$(1 - t)(f(x, u) - f(0, u)) + t(g(x, u) - g(0, u)).$$

Our claim means that we have to be able to find $\Phi \in \mathcal{E}(n+r+1, n+r)$ as a germ at the point $(0, 0, t_0)$ and $\alpha \in \mathcal{E}(r + 1)$ at the point $(0, t_0)$ such that $\Phi(x, u, t) = \Phi_t(x, u)$ has the form $(\phi_t(x, u), \psi_t(u)) \in \mathbf{R}^n \times \mathbf{R}^r$ and further, if $\alpha_t(u) = \alpha(u, t)$, then

(a) $\Phi_{t_0} = \mathrm{id} \in \mathcal{B}(n + r), \quad \alpha_{t_0} = 0,$

(b) $\Phi_t | \mathbf{R}^n \times \{0\} = \mathrm{id} \in \mathcal{B}(n), \quad \alpha_t(0) = 0,$

(c) $F_t \circ \Phi_t + \alpha_t = F_{t_0}.$

These conditions state that (Φ_t, α_t) is a morphism between (r, F_{t_0}) and (r, F_t) which is an isomorphism by (a).

Because of (a), we may replace (c) by the differential condition $\frac{\partial}{\partial t}(F_t \circ \Phi_t + \alpha_t) = 0$, which may be expanded to

$$\text{(d)} \quad \sum_{i=1}^{n} \frac{\partial F}{\partial x_i}(\Phi, t) \frac{\partial \phi_i}{\partial t}(x, u, t) + \sum_{j=1}^{r} \frac{\partial F}{\partial u_j}(\Phi, t) \frac{\partial \psi_j}{\partial t}(u, t)$$

$$+ \frac{\partial F}{\partial t}(\Phi, t) + \frac{\partial \alpha}{\partial t}(u, t) = 0.$$

Note that (Φ, t) should really be written $(\Phi(x, u, t), t)$.

So, now we have replaced (c) by (d) and we must try to 'solve' the conditions for $\partial\phi/\partial t$, $\partial\psi/\partial t$, $\partial\alpha/\partial t$. We are looking for germs

$$\xi_i \in \mathcal{E}(n + r + 1), \quad i = 1, \ldots, n,$$

$$\zeta_j \in \mathcal{E}(r + 1), \qquad j = 1, \ldots, r + 1,$$

which satisfy

$$\text{(e)} \quad \sum_i \frac{\partial F}{\partial x_i} \xi_i + \sum_j \frac{\partial F}{\partial u_j} \zeta_j + \zeta_{r+1} = - \frac{\partial F}{\partial t} \in \mathcal{E}(n + r + 1)$$

$$\xi_i | \mathbf{R}^n \times \{0\} \times \mathbf{R} = 0, \quad \text{i.e.,} \ \xi_i \in \mathfrak{m}(r). \mathcal{E}(n + r + 1),$$

$$\zeta_j | \{0\} \times \mathbf{R} = 0, \quad \text{i.e.,} \ \zeta_j \in \mathfrak{m}(r). \mathcal{E}(r + 1).$$

To see that this is what we want, suppose that the ξ_i and ζ_j have been found. Then if Φ and α are solutions of the differential equations

$$\partial\phi_i/\partial t = \xi_i(\phi, \psi, t),$$

$$\partial\psi_j/\partial t = \zeta_j(\psi, t), \quad j \leq r,$$

$$\partial\alpha/\partial t = \zeta_{r+1}(\psi, t),$$

with initial conditions $\Phi_{t_0} = \mathrm{id}$, $\alpha_{t_0} = 0$, then Φ and α satisfy (a), (b) and (d).

Now, because $\partial F/\partial t = g - f \in \mathfrak{m}(r)\mathcal{E}(n + r + 1)$, it is sufficient for the proof of (e) to demonstrate that

$$\mathfrak{m}(r)\mathcal{E}(n+r+1) \subset \langle \partial F/\partial x_i \rangle_{\mathfrak{m}(r)\mathcal{E}(n+r+1)} + \langle \partial F/\partial u_j, 1 \rangle_{\mathfrak{m}(r)\mathcal{E}(r+1)}$$

where $\langle b_1, \ldots, b_k \rangle_A$ is defined to be $\{\sum_i a_i b_i \mid a_i \in A\}$, as usual.

For this inclusion we only need to show that

$$(*) \qquad \mathcal{E}(n+r+1) = \langle \partial F/\partial x_i \rangle_{\mathcal{E}(n+r+1)} + \langle \partial F/\partial u_j \rangle_{\mathcal{E}(r+1)} + \mathcal{E}(r+1).$$

The fact which we want to use is that F_t is a k-transversal unfolding of η. By lemma 16. 4, we have

$$\mathfrak{m}(n) = \langle \partial\eta/\partial x_i \rangle_{\mathcal{E}(n)} + \langle \partial F_t/\partial u_j | \mathbf{R}^n \times \{0\} \rangle_{\mathbf{R}} + \mathfrak{m}(n)^{k+1}.$$

The middle term on the right is simplified because we have assumed that $F_t | \{0\} \times \mathbf{R}^r = 0$. Since η is k-determined, i. e. $\mathfrak{m}(n)^{k+1} \subset \langle \partial\eta/\partial x_i \rangle$, we may omit the last term. Substituting $\partial\eta/\partial x_i = \partial F_t/\partial x_i | \mathbf{R}^n \times \{0\}$ gives

$$\mathfrak{m}(n) = \langle \partial F_t/\partial x_i | \mathbf{R}^n \times \{0\} \rangle_{\mathcal{E}(n)} + \langle \partial F_t/\partial u_j | \mathbf{R}^n \times \{0\} \rangle_{\mathbf{R}}.$$

To prove (*) we examine the equation

$$\mathcal{E}(n + r + 1) = \mathfrak{m}(n)\mathcal{E}(n + r + 1) + \mathcal{E}(r + 1).$$

If $g \in \mathfrak{m}(n)\mathcal{E}(n + r + 1)$ then we have just shown that there is an element in

$$\langle \partial F/\partial x_i \rangle_{\mathcal{E}(n+r+1)} + \langle \partial F/\partial u_j \rangle_{\mathcal{E}(r+1)}$$

which agrees with g on $\mathbf{R}^n \times \{0\} \times \{t_0\}$, at least. The elements of $\mathcal{E}(n + r + 1)$ which vanish on $\mathbf{R}^n \times \{0\} \times \{t_0\}$ lie in $\mathfrak{m}(r+1). \mathcal{E}(n+r+1)$. Putting things together we have

$$(**) \qquad \langle \frac{\partial F}{\partial x_i} \rangle_{\mathcal{E}(n+r+1)} + \langle \frac{\partial F}{\partial u_j} \rangle_{\mathcal{E}(r+1)} + \mathcal{E}(r+1) + \mathfrak{m}(r+1)\mathcal{E}(n+r+1) = \mathcal{E}(n+r+1).$$

Let $C = \mathcal{E}(n + r + 1)$, thought of as a finitely generated $\mathcal{E}(n+r+1)$-module,

$$A = \langle \partial F/\partial x_i \rangle_{\mathcal{E}(n+r+1)} \quad \text{which is a submodule of } C, \text{ and}$$

$$B = \langle \partial F/\partial u_j \rangle_{\mathcal{E}(r+1)} + \mathcal{E}(r + 1).$$

As $\mathcal{E}(r+1)$-modules, $B \subset C$, where C is an $\mathcal{E}(r+1)$-module via the inclusion $\mathcal{E}(r + 1) \subset \mathcal{E}(n + r + 1)$. Note B is finitely generated over $\mathcal{E}(r + 1)$. We know

$$(**) : A + B + \mathfrak{m}(r + 1)C = C$$

and want

$$(*) \quad : A + B = C.$$

To deduce this from the given information about A, B and C we may first put $A = 0$ (compute modulo A). So we consider the map $\mathcal{E}(r + 1) \to \mathcal{E}(n + r + 1)$ which is induced by projection and which makes each $\mathcal{E}(n+r+1)$-module into an $\mathcal{E}(r+1)$-module. For our modules we know that

B is finitely generated over $\mathcal{E}(r + 1)$,
C is finitely generated over $\mathcal{E}(n + r + 1)$,
$B + \mathfrak{m}(r + 1)C = C$.

The generators $b_1, \ldots, b_s$ of B generate the vector space $C/\mathfrak{m}(r + 1)C$. Hence by corollary 6.6 to the preparation theorem they also generate C as an $\mathcal{E}(r+1)$-module. Therefore $B = C$.

Every step in the proof of Mather's theorem on universal unfoldings has now been completed. $\checkmark$

17 · Pictures of the seven elementary catastrophes

Literature: A. N. Godwin: Three dimensional pictures for Thom's parabolic umbilic, I. H. E. S. Publ. Math., 40 (1971), 117-38.
G. Wassermann: Stability of unfoldings, Dissertation, Regensburg 1973, Springer Lecture Notes, 393 (1974).
G. Wassermann: (r, s)-stability of unfoldings, preprint.

Recall that the theory of elementary catastrophes is local in nature. There is a family of potential functions $V_u : X \to \mathbf{R}$ where X is a subset of $\mathbf{R}^n$, containing a neighbourhood of the origin, and the parameters u lie in an open subset $U \subset \mathbf{R}^r$. We may assume $X = \mathbf{R}^n$. A particular catastrophe determines a germ $\eta \in m(n)^2$ which is unfolded to a germ (r, f), $f \in m(n + r)$.

The coordinates in $\mathbf{R}^r$ are both the unfolding parameters of η and the external parameters of the model.

Definitions.

A local regime at $u \in U$ is any one of the local minima of $f|\mathbf{R}^n \times \{u\}$.

A process (simplest interpretation) for the germ η (or f) is a section s of the bundle $\bar{\mathbf{R}}^n \times U \to U$, such that $(s(u), u)$ is either a local regime or at infinity. The section should be defined on an open, dense subset.

A convention assigns a process to the unfolding f.

A regular point of a process is a point in the subset U where the section s is locally defined and continuous on a neighbourhood of the point. This is equivalent to saying that over a neighbourhood of the point a homeomorphism of the bundle exists taking the section s onto a constant section.

A catastrophe point is a non-regular point in U.

The morphology of the catastrophe is the set of all catastrophe points, one also refers to catastrophe set. To study the geometry of a singularity η with its unfolding f the first important subset is

$$\Sigma_f = \{(x, u) \in \mathbf{R}^n \times U \mid d_x f(x, u) = 0\}$$

where $d_x f(-, u) = D(f \mid \mathbf{R}^n \times \{u\})$.

For each fixed u, the points of Σ_f lying in $\mathbf{R} \times \{u\}$ give the local minima (regimes) and maxima of f at u. Next consider

$$\Delta_f = \{(x, u) \in \Sigma_f \mid d_x^2 f(x, u) \text{ is degenerate}\}$$

and its projection $D_f = \pi(\Delta_f)$ under the projection $\pi : \mathbf{R}^n \times U \to U$. The points of D_f are important candidates for catastrophe points.

As to conventions there are two basic ones to consider. The Maxwell convention states that s(u) is at the point where $f \mid \mathbf{R}^n \times \{u\}$ has its lowest minimum. As this may be at $-\infty$ this convention is best used when f has only finite minima. Clearly catastrophe points occur when $f \mid \mathbf{R}^n \times \{u\}$ attains an absolute minimum in two places.

With U now a subset of space-time, the next convention requires U to be foliated by non-singular 1-dimensional leaves, each determined by a fixed position in space and parametrised by time. The axes of the local model do not need to be the local cartesian coordinates of space-time.

The perfect-delay convention states that the section s will remain continuous for as long as possible. This means that along the leaves of time, s(u) will follow a continuous family of minima until these minima disappear. Only then will s jump to another family of minima.

There are more refined concepts of process and it is very common in the work of Zeeman to find that a process assigns to each smooth path τ in U a section s_τ of the bundle $\mathbf{R}^n \times \tau \to \tau$ such that $(s_\tau(u), u)$ is either a local regime or at infinity. In such a scheme each path has a direction in which the parameter increases and the perfect delay convention operates along each path: s_τ remains continuous for as long as possible along the path. This description does not do justice to Zeeman's models, but then the previous explanation doesn't do justice to Thom's.

Now we describe the seven catastrophes

The fold

Unfolding: $f(x, u) = x^3 + ux$

$$\Sigma_f = \{(x, u) \mid 3x^2 + u = 0\}$$

$$\Delta_f = \{(0, 0)\}$$

$$D_f = \{0\}$$

Here we may draw the graph of f. For u negative, f has one local minimum, for u positive f has none (diagram 1).

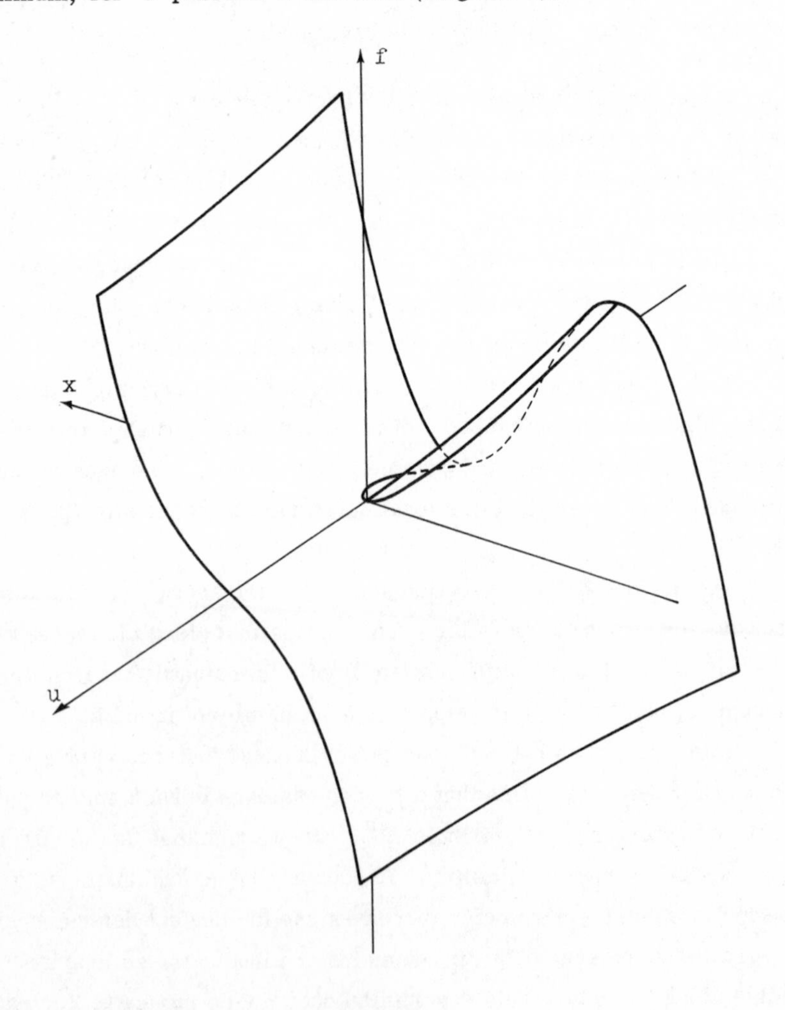

Diagram 1

The picture of Σ_f in the x, u-plane is given in diagram 2.

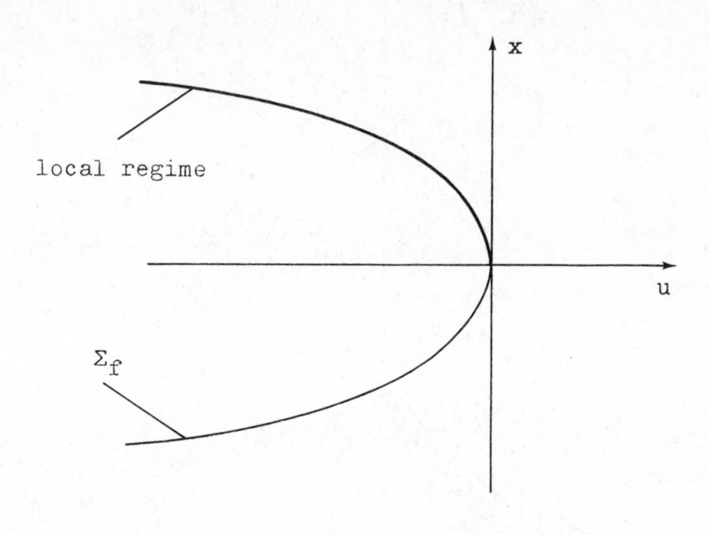

Diagram 2

Clearly for positive u, the process s can only be at infinity; for u negative, $s(u) = -\sqrt{-u/3}$ or at infinity. Thus if u is a time coordinate one may imagine a phenomenon, described by a parameter x, which exists up to time $u = 0$ in a state given by $x = -\sqrt{-u/3}$. At time $u = 0$ the system jumps to another state not described by this local model (this does not imply that the phenomenon disappears altogether).

A significant feature of the model is the rapid change of the parameter x just before the disappearance of the local regime.

The cusp

Unfolding: $f(x, u, v) = x^4 - ux^2 + vx$

$$\Sigma_f = \{(x, u, v) \mid 4x^3 - 2ux + v = 0\}$$
$$\Delta_f = \{(x, u, v) \in \Sigma_f \mid 12x^2 - 2u = 0\}$$
$$D_f = \{(u, v) \mid 27v^2 = 8u^3\}$$

Here we may draw Σ_f (diagram 3 - after Zeeman).

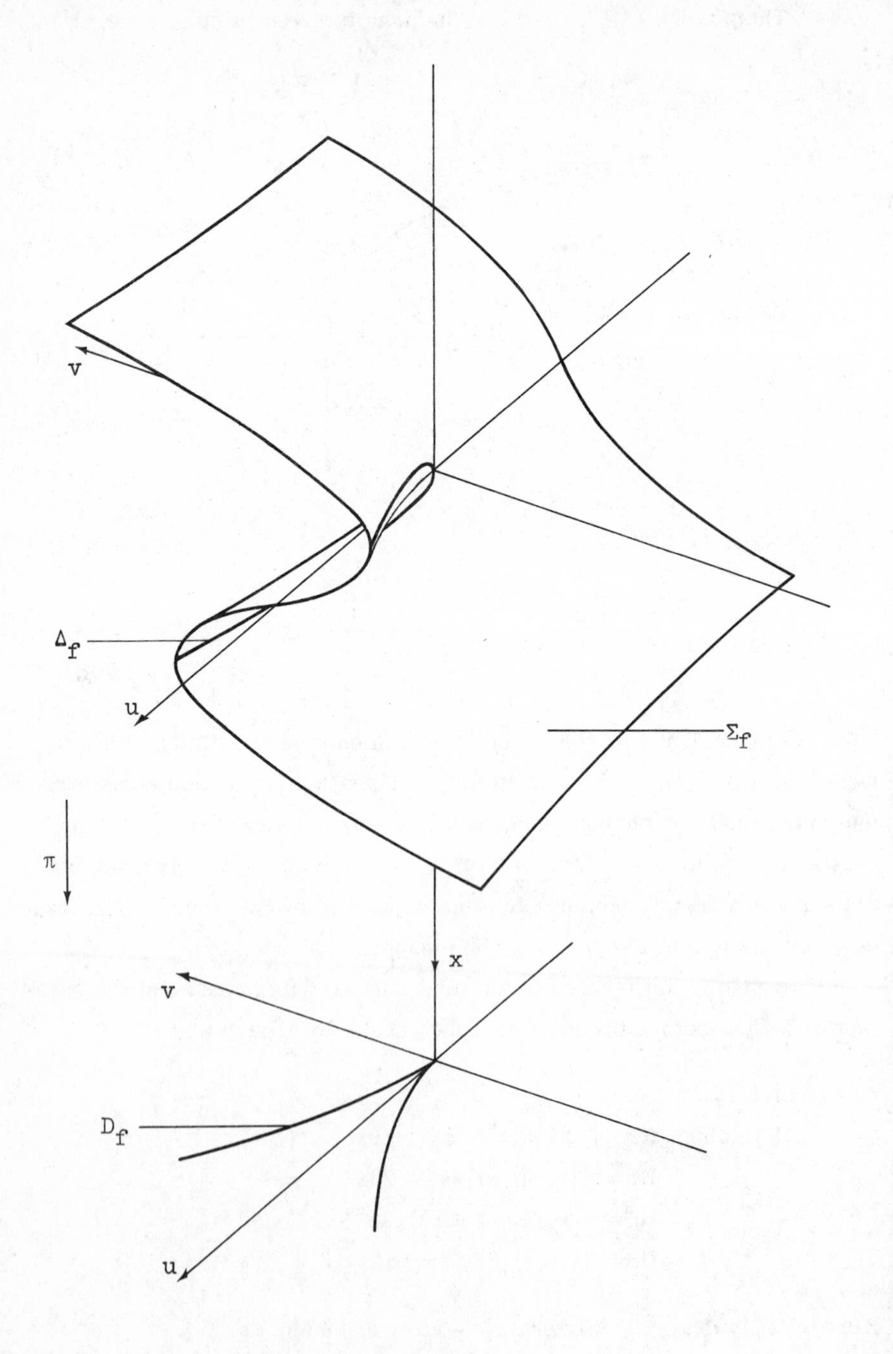

Diagram 3

To each point of the u, v-plane there corresponds a graph of f depending only on x. The various possibilities appear in diagram 4 (replacing v by -v reflects the small pictures in the vertical axis).

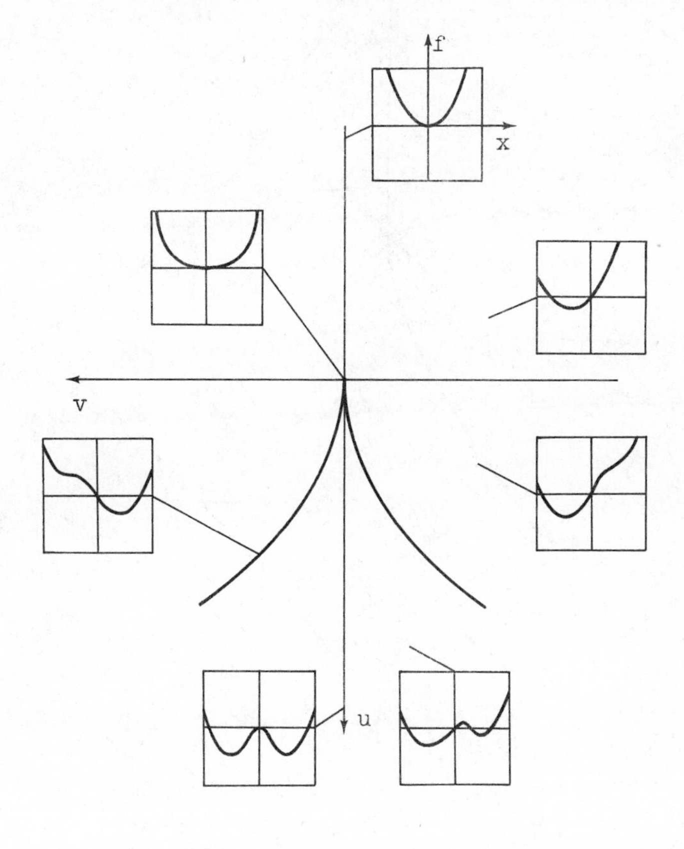

Diagram 4

In diagram 4, the positive u-axis is the Maxwell set where there is a switch in which minimum is lower. Minima disappear across the lines of the cusp. With the Maxwell convention the process has the graph shown in diagram 5.

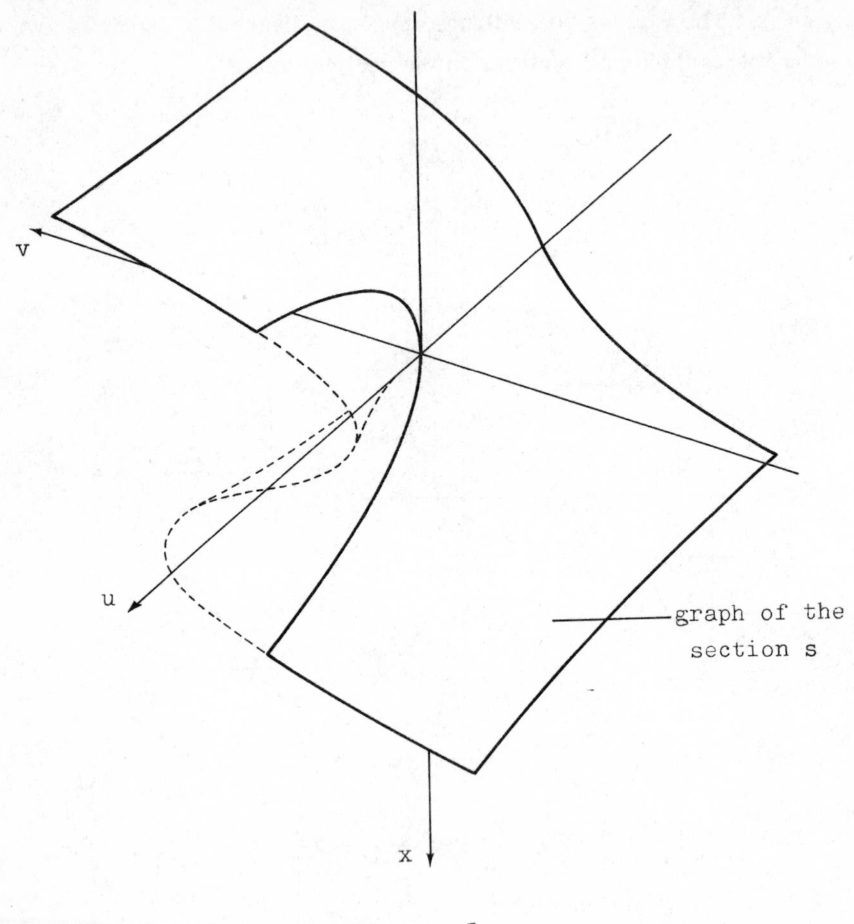

Diagram 5

A possible process with perfect-delay is shown in diagram 6. The catastrophe set follows the cusp line until the time-leaves become tangent to the curve.

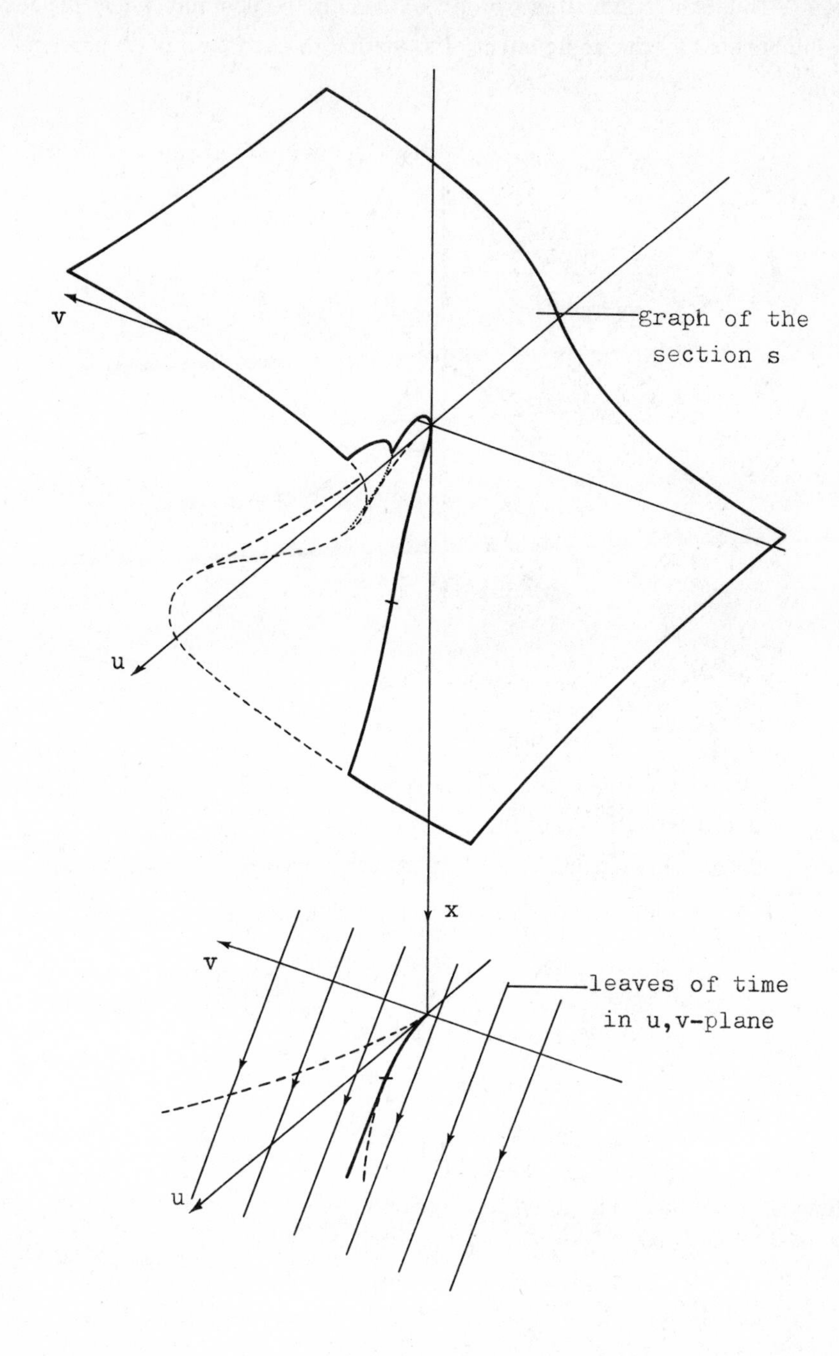

Diagram 6

Diagram 7 indicates two models which are possible when 'process' is interpreted as the assignment of a section to each path (with perfect-delay).

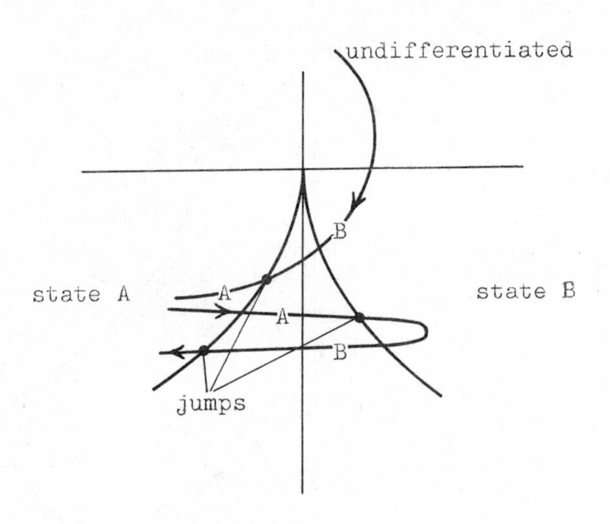

Diagram 7

The dove-tail (swallow-tail)

Unfolding: $f(x, u, v, w) = x^5 + ux^3 + vx^2 + wx$

$$\Sigma_f = \{(x, u, v, w) \mid 5x^4 + 3ux^2 + 2vx + w = 0\}$$

$$\Delta_f = \{(x, u, v, w) \in \Sigma_f \mid 20x^3 + 6ux + 2v = 0\}$$

$$D_f = \{(u, v, w) \mid \exists x \text{ s.t. } 5x^4 + 3ux^2 + 2vx + w = 0 \text{ and } 20x^3 + 6ux + 2v = 0\}$$

Here D_f has the form given in diagram 8.

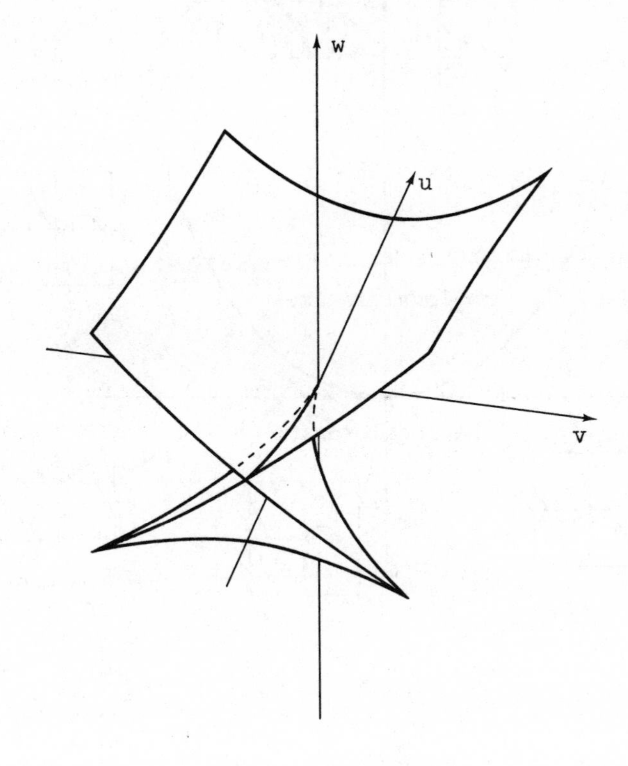

Diagram 8

For positive and negative u the distribution of regimes is given in diagrams 9 and 10.

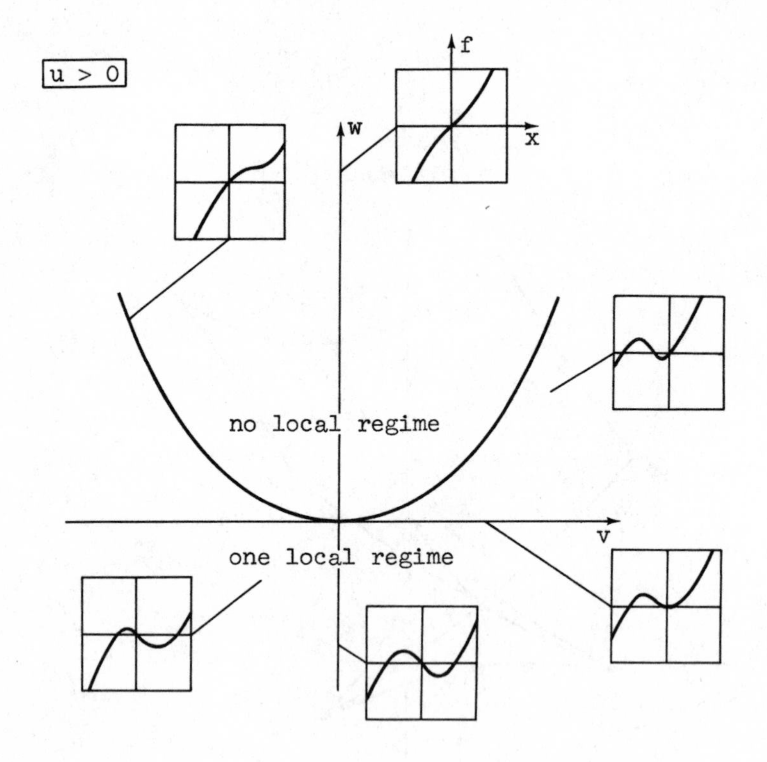

Diagram 9

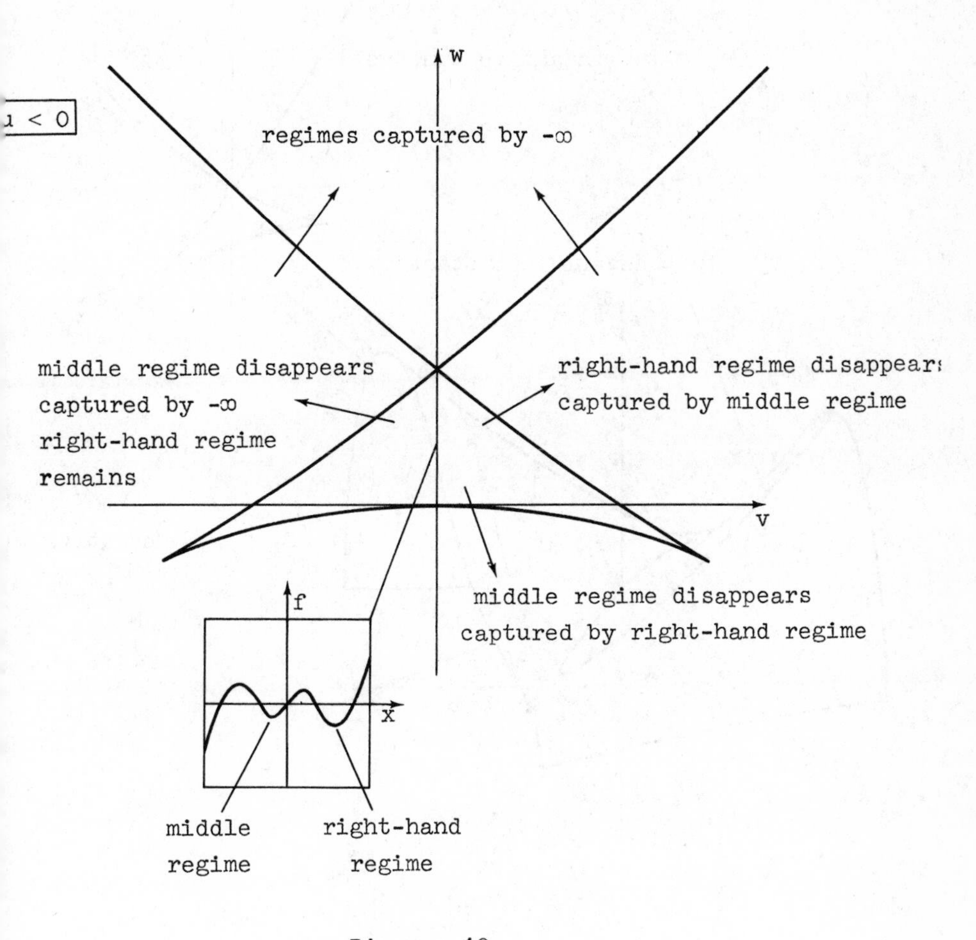

Diagram 10

For negative u, there is an illuminating picture in (x, v, w) space: diagram 11.

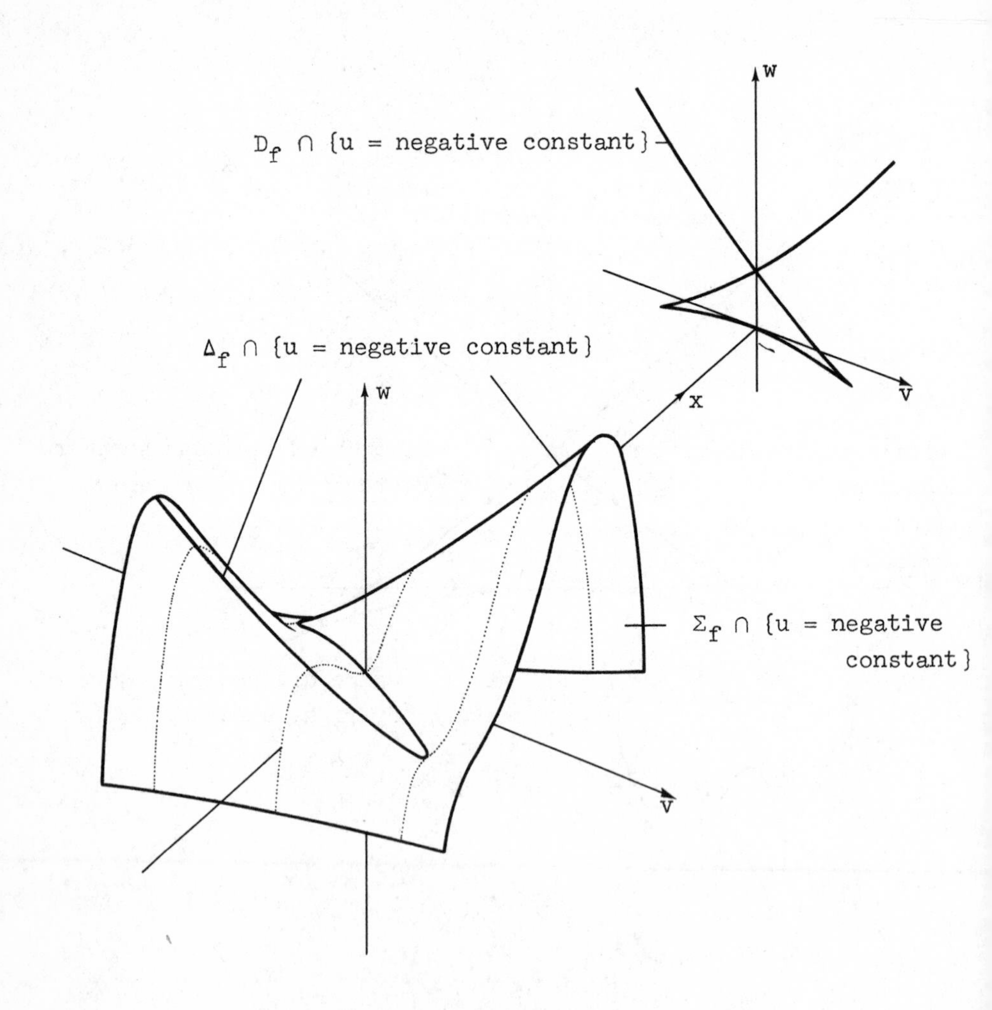

Diagram 11

If time flows parallel to the w-axis, the perfect-delay convention gives the catastrophe set shown in diagram 12. We assume that the phenomenon begins by being governed by the local regime and not $-\infty$.

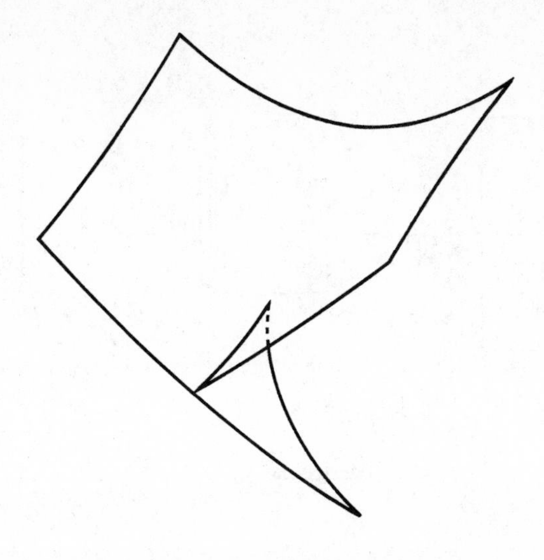

Diagram 12

<u>The hyperbolic umbilic</u>

Unfolding: $f(x, y, u, v, w) = x^3 + y^3 + wxy - ux - vy$

$$\Sigma_f = \{(x, y, u, v, w) \mid 3x^2 + wy - u = 3y^2 + wx - v = 0\}$$

$$\Delta_f = \{(x, y, u, v, w) \in \Sigma_f \mid \det\begin{bmatrix} 6x & w \\ w & 6y \end{bmatrix} = 0\}$$

$$D_f = \{(u, v, w) \mid \exists (x, y) \text{ with } u = 3x^2 + wy$$

$$v = 3y^2 + wx,\ w^2 = 36xy\}$$

For each fixed w, Σ_f determines a map $\mathbf{R}^2 \to \mathbf{R}^2$: $(x, y) \mapsto (u, v) = (3x^2 + wy, 3y^2 + wx)$ and the intersection of D_f with the plane $\{w = \text{constant}\}$ is the set of singular values of this map.

For $w = 0$, the map is a folded handkerchief (diagram 13) and for $w \neq 0$ this becomes more generic, diagram 14.

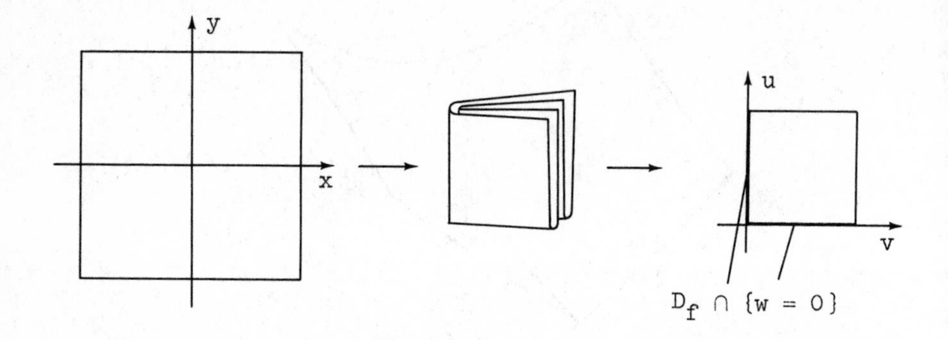

Diagram 13

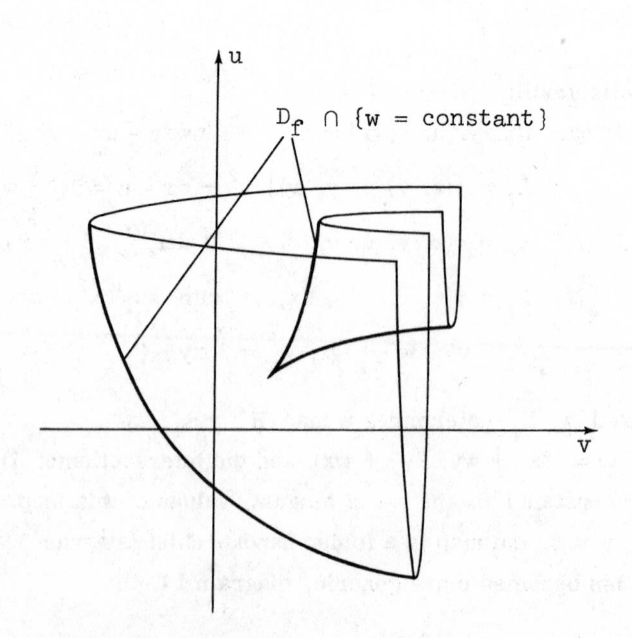

Diagram 14

In u, v, w-space D_f is symmetrical: by reflection in the u, v-plane. Half of D_f is drawn in diagram 15.

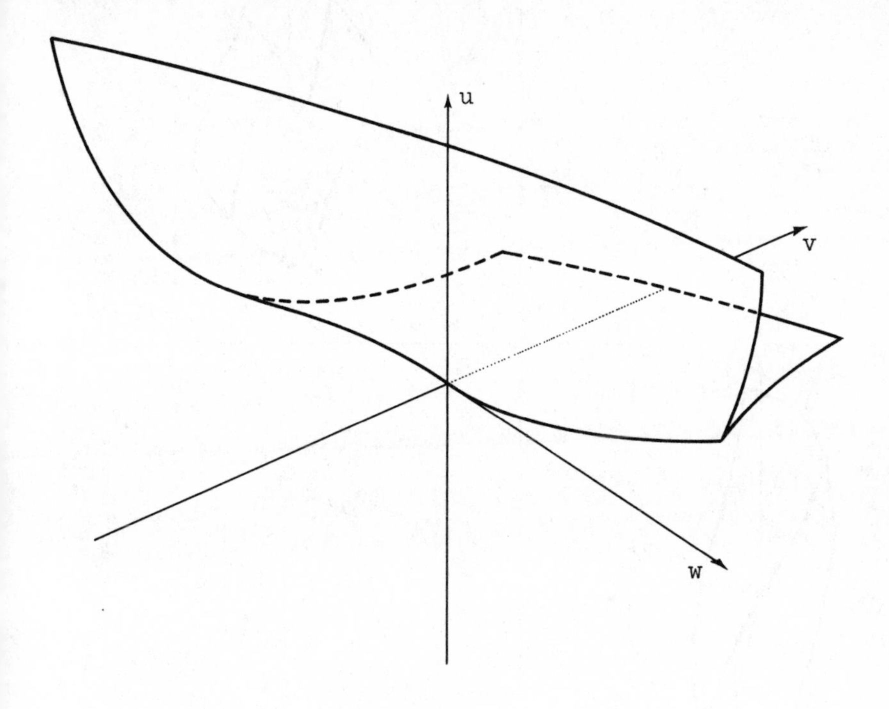

Diagram 15

For $w = 0$ we indicate the graph of f for $u = v =$ positive and $u = v =$ negative, diagram 16. For $w \neq 0$ the pictures do not change very much. Closer study reveals that there is always at most one finite local regime and this occurs on that side of the surface in diagram 15 which contains the quadrant $w = 0$, $u, v > 0$.

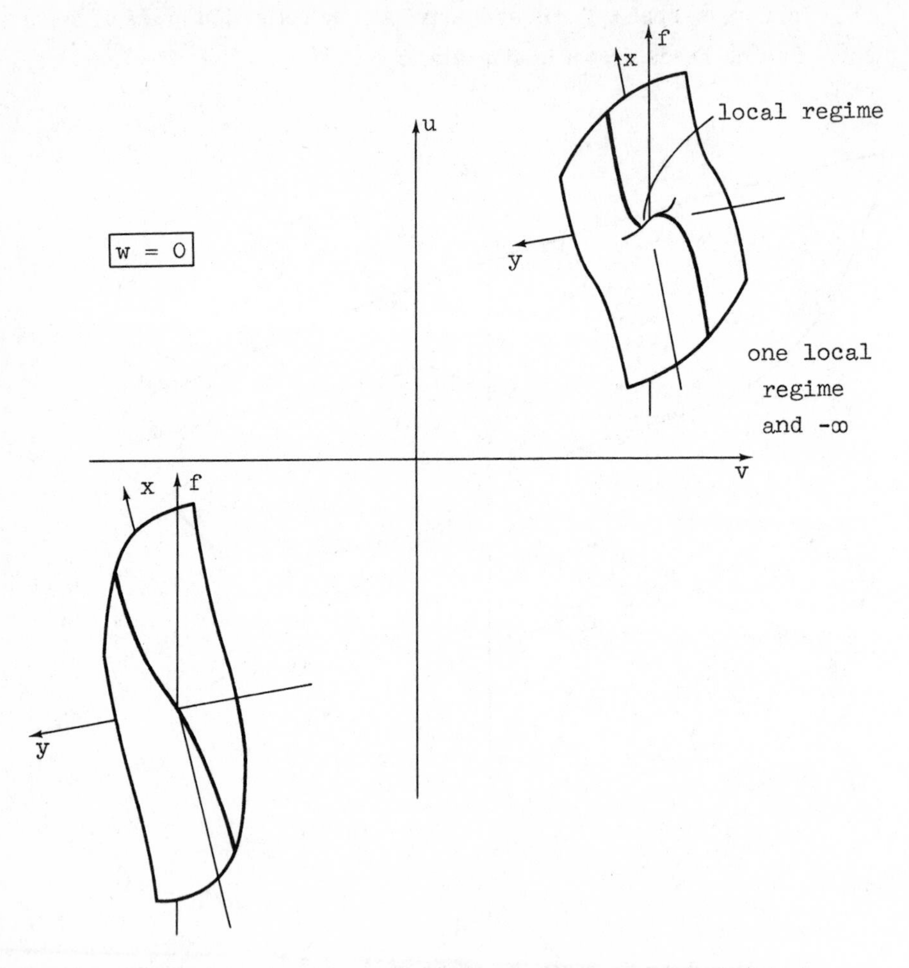

Diagram 16

<u>The elliptic umbilic</u>

Unfolding: $f(x, y, u, v, w) = \frac{x^3}{3} - xy^2 + w(x^2 + y^2) - ux - vy.$
This change of coordinates simplifies the form of D_f:

$$D_f = \{(u, v, w) : \exists x, y \text{ s.t. } u = x^2 - y^2 + 2wx$$

$$v = -2xy + 2wy, \ x^2 + y^2 = w^2\}$$

If we consider the coordinates $z = x + iy$ and $u + iv$ in **C** we see that for fixed w, $D_f \cap \{w = \text{constant}\}$ is the image of the circle $|z| = |w|$ under the map $(x, y) \mapsto (u, v) = (x^2 + y^2 + 2wx, 2wy - 2xy)$, that is $z \to \bar{z}^2 + 2wz$.

Taking $w = 1$ we obtain the curve $2e^{i\theta} + e^{-2i\theta}$, diagram 17.

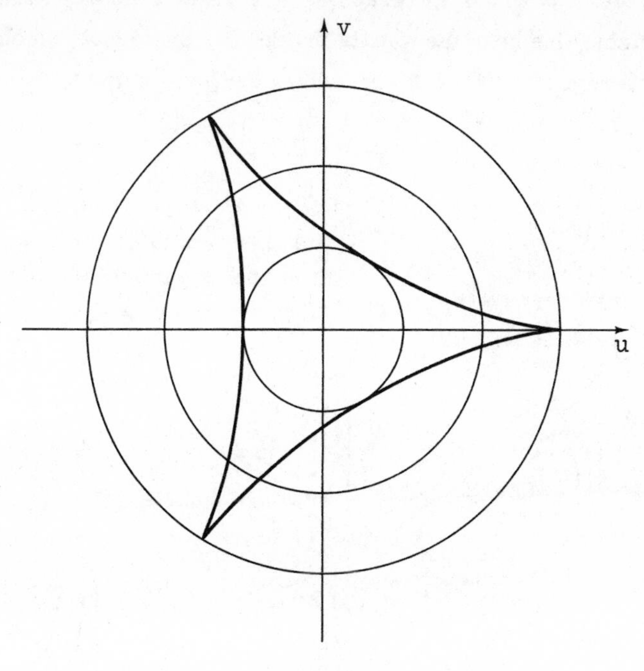

Diagram 17

Hence D_f is the set shown in diagram 18.

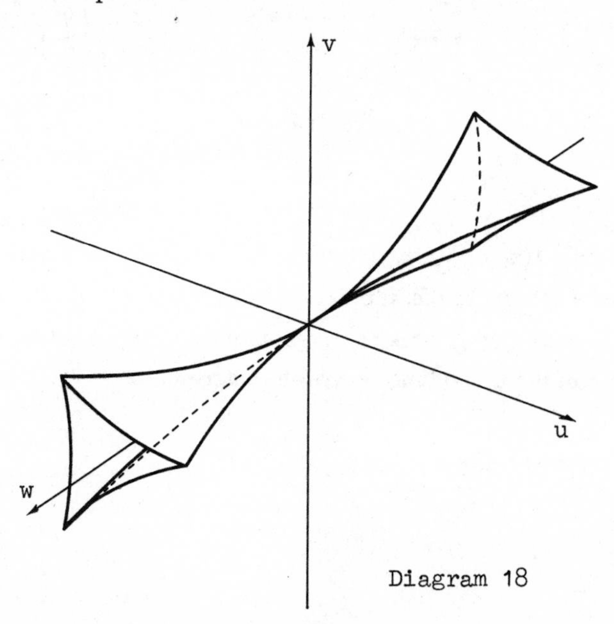

Diagram 18

One finds only one local regime, and that occurs inside D_f, for $w > 0$. At $u = v = w = 0$ the graph of $f(x, y)$ is a monkey saddle. Along a suitable line near the w-axis inside D_f, this saddle unfolds as shown in diagram 19. The diagram shows the level-curves.

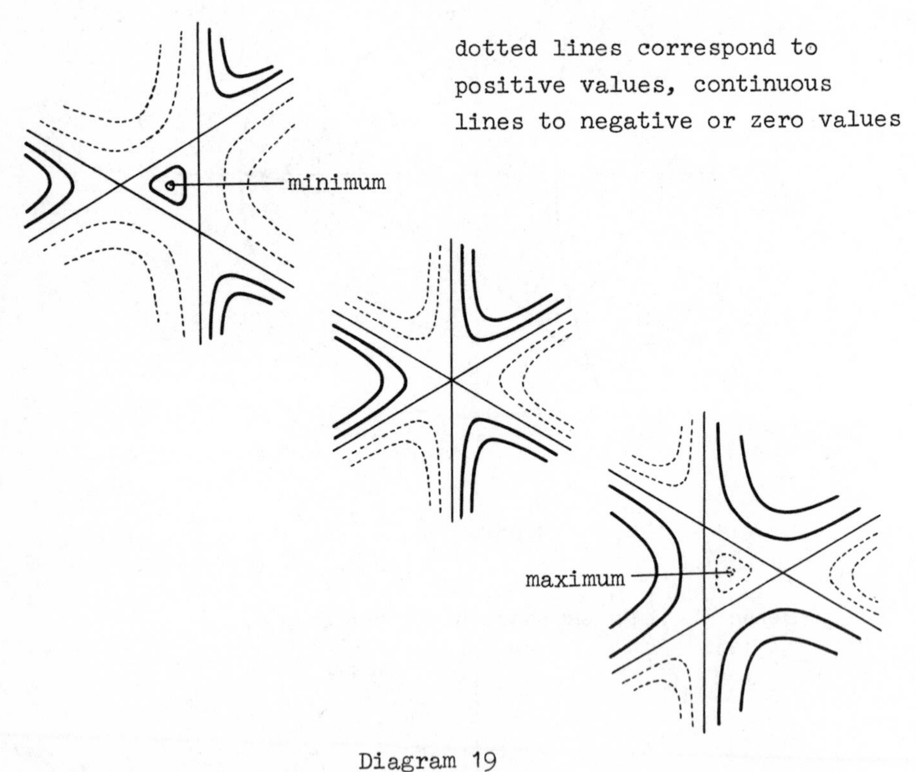

Diagram 19

The Butterfly

Unfolding $f(x, u, v, w, t) = x^6 + tx^4 + ux^3 + vx^2 + wx$. Here we draw a 'clock' of the set D_f. Taking fixed (u, t), the intersection of D_f with the v, w-plane gives a curve. As we move around the unit circle in the u, t-plane we obtain diagram 20.

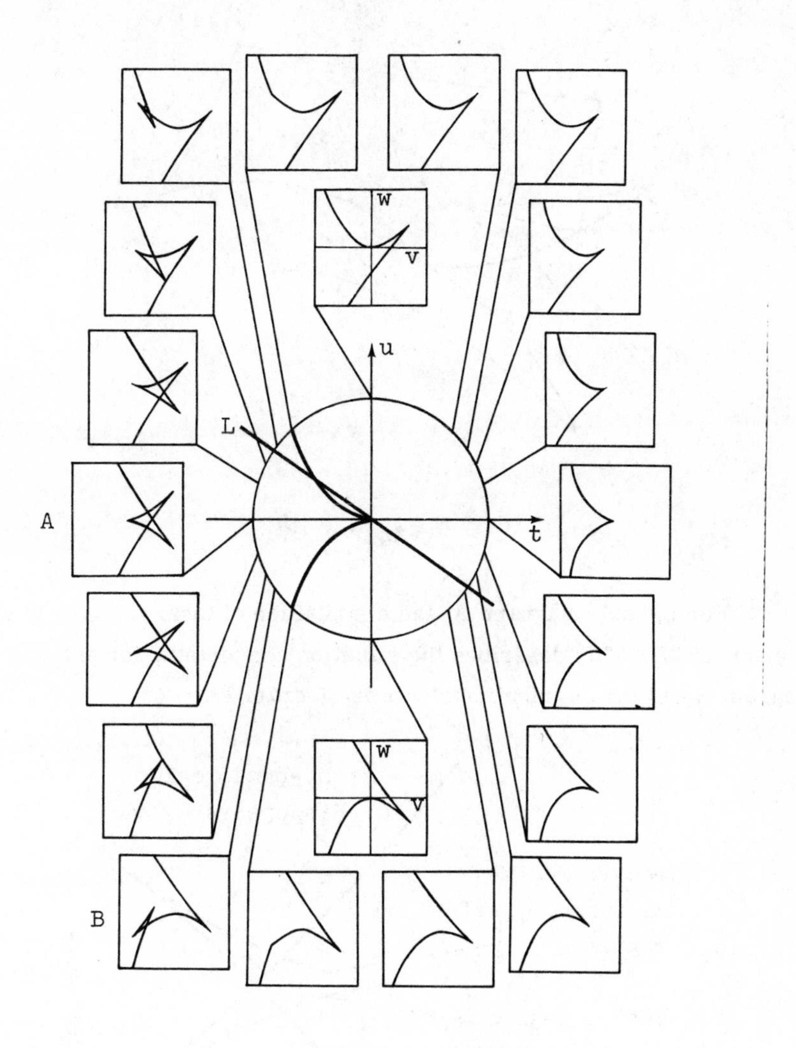

Diagram 20

The cuspidal curve is the dove-tail (swallowtail) line where the formation of the dovetail catastrophe occurs. The surface of D_f generated along the line L is shown in diagram 21.

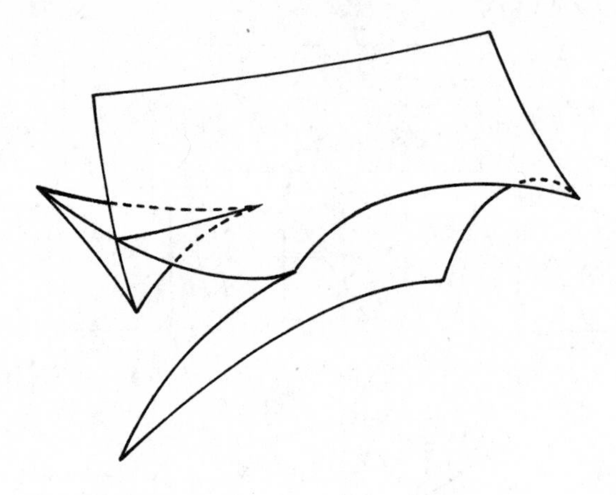

Diagram 21

For the sub-diagram A the distribution of local regimes is shown in diagram 22. This describes the situation completely, for example sub-diagram B has the distribution shown in diagram 23.

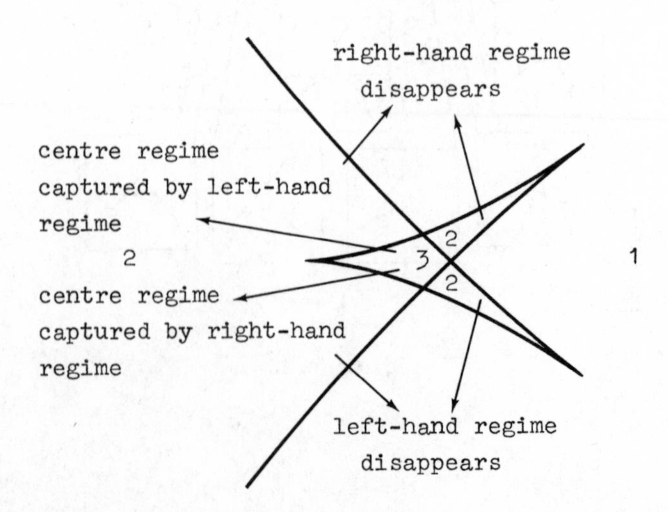

Diagram 22

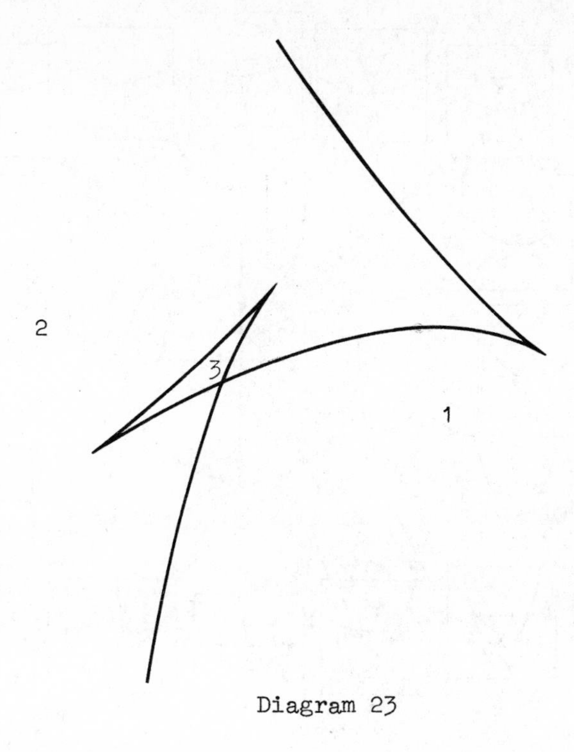

Diagram 23

The parabolic umbilic

Unfolding $f(x, y, u, v, w, t) = x^2y + y^4 + wx^2 + ty^2 - ux - vy$.
The 'clock' is shown in diagram 24. This picture seems to have originated from Chenciner. A detailed study with a great deal of further information and many pictures is to be found in Godwin's paper.

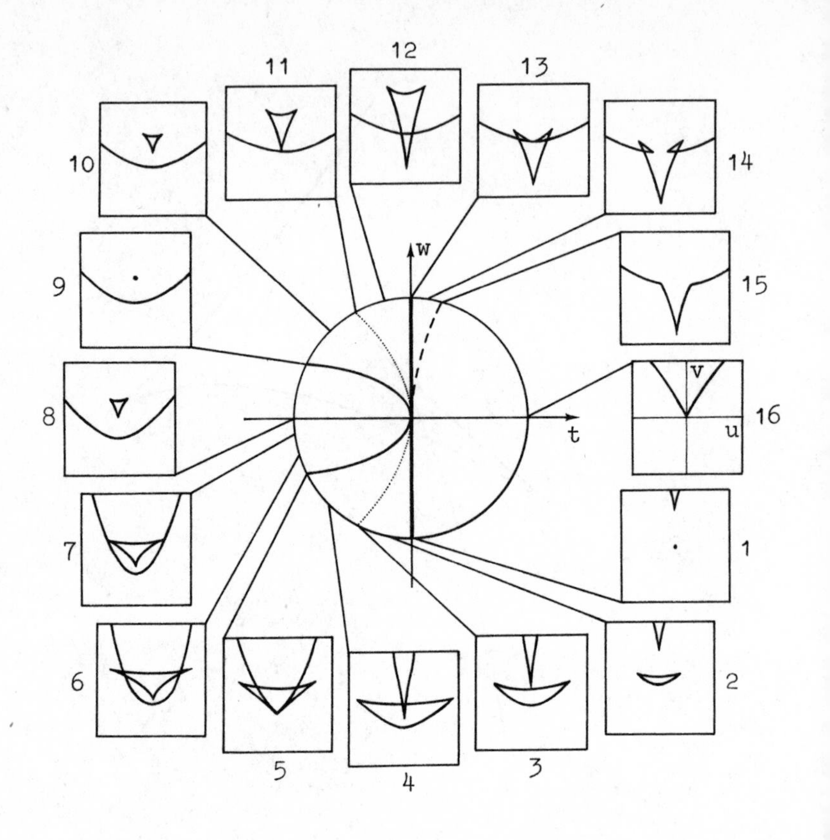
11
12
13
10
14
w
9
15
v
8
16
t
u
7
1
6
2
5
4
3

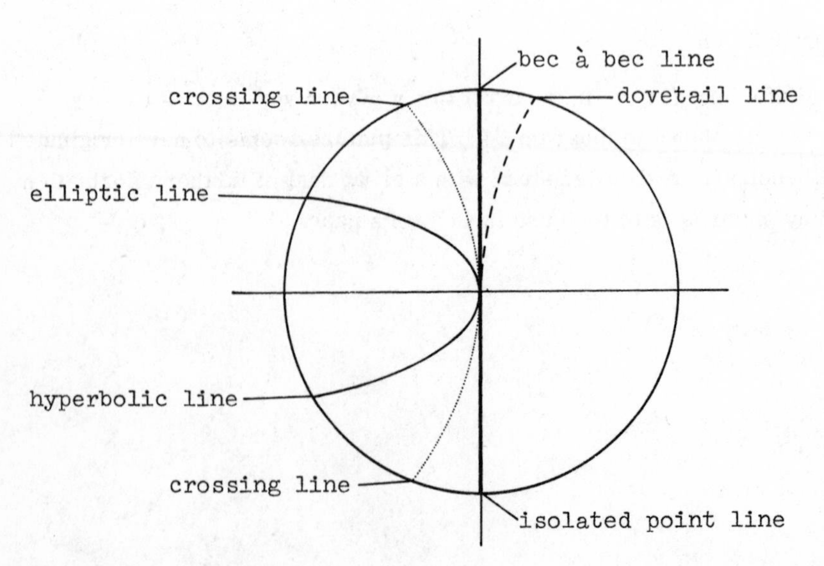
bec à bec line
crossing line
dovetail line
elliptic line
hyperbolic line
crossing line
isolated point line

Diagram 24

The explanation of the various curves in the w, t-plane is the following:

isolated point	line	: an isolated point appears
crossing	line	: distinct parts of D_f meet each other
hyperbolic	line	: centre of a hyperbolic umbilic
elliptic	line	: centre of an elliptic umbilic
bec à bec	line	: breaking apart of the cusps of two dovetails (swallowtails)
dovetail	line	: centre of the appearance of two dovetails (swallowtails).

The number of local regimes is indicated in diagram 25.

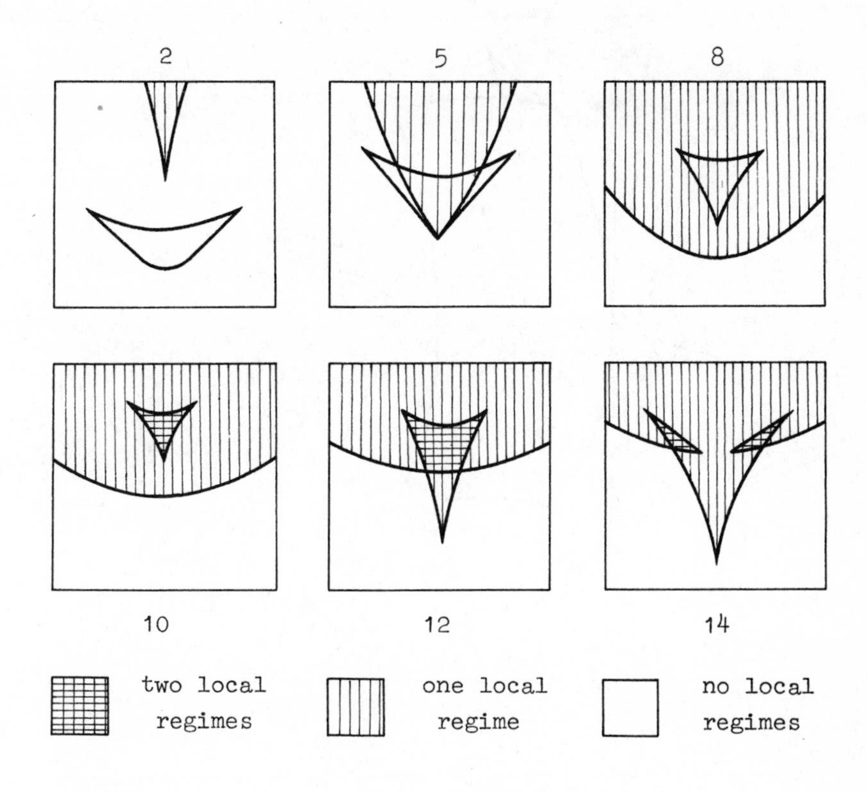

Diagram 25

Unfoldings in higher dimensions

We have drawn pictures of minimum-dimensional unfoldings. But $f(x, u, v, w) = x^4 - ux^2 - vx$ is a versal unfolding and the sets Σ, Δ and D are just suspensions of those for the ordinary cusp catastrophe (by taking the products with the w-axis). In fact, any suspension of a universal unfolding is versal. The cusp given above is equivalent to the following one:

$$f(x, u, v, w) = x^4 - (u - w^2)x^2 - vx,$$

and D_f has the form shown in diagram 26. If u represents time then planes of constant time cut D_f as indicated. There is probably a convention which has D_f as catastrophe set.

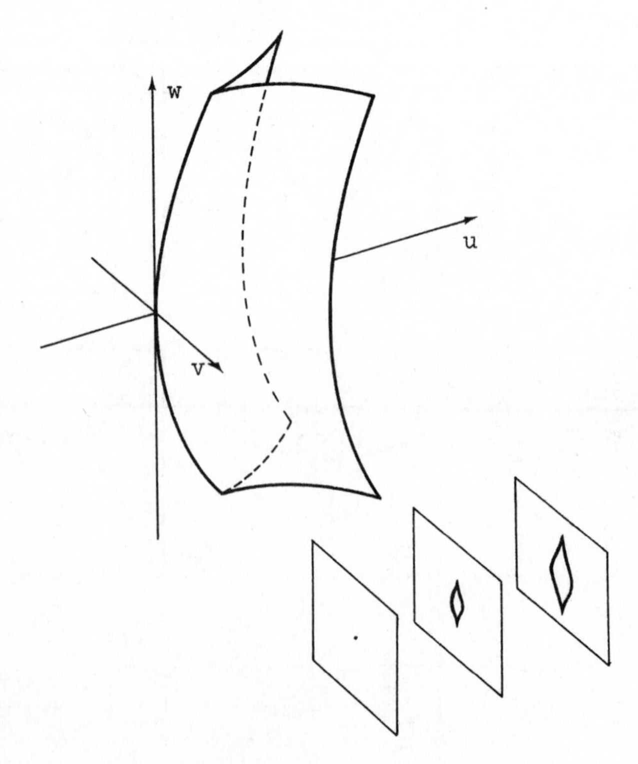

Diagram 26

Time-stability

The elementary theory discussed above depends on the classification of stable germs up to equivalence. The equivalence is defined by means of changes of coordinates on the right and on the left. Details are given in Wassermann's dissertation.

One significant feature is that germs of arbitrary diffeomorphisms of U are permitted. Wassermann (second reference) has considered another problem where the diffeomorphisms of U are restricted. One supposes U to be foliated by subspaces of constant time - there are coordinates on U where one of the axes represents time. One insists that any change of coordinates in U maps leaves of constant time onto leaves of constant time. Maps which are equivalent in this restricted sense are called time-equivalent or t-equivalent, and the stable ones are called t-stable. A classification of t-stable functions is now possible. The list is more involved than that for elementary catastrophes but it is still finite for low dimensions. One t-stable unfolding of the cusp is

$$f(x, u, v, t) = x^4 + ux^2 + tx + ux + v^2x.$$

The time axis naturally has coordinate t. An unfolding like that for diagram 26 with the time running along the u-axis and tangent to the cusp point would not be t-stable.

A picture of D_f is given in diagram 27. The intersection with planes of constant time is also shown.

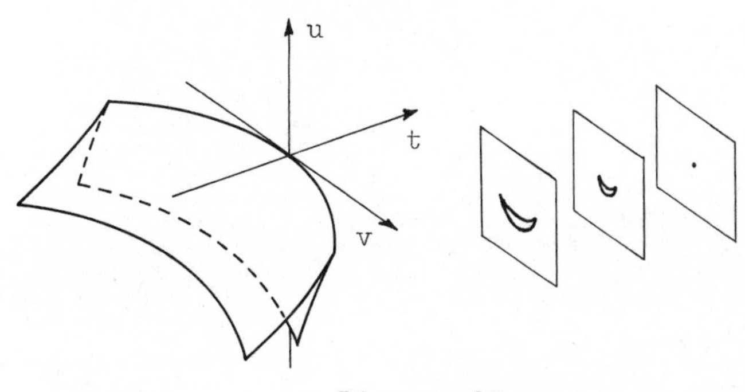

Diagram 27

"The cat catches the mouse."

'. . . . For an animal, feeding, that is restoring its reserves in chemical energy, is the most fundamental regulative process. This also is a periodic process, hence it is described by a loop, which we call the predation loop.

Here we meet with a fundamental difficulty: predation implies the presence of a prey, that is a being external to the animal itself. Feeding is - fundamentally - engulfing a prey in the organism (as seen very clearly in phagocytosis for the Unicellulars). Hence we have, to describe the predation loop, to use the simplest of the capture catastrophes, the Riemann-Hugoniot catastrophe. The predation loop is the unit circle in the Ouv plane of the unfolding $V = x^4/4 + ux^2/2 + vx$. This circle meets the bifurcation curve $4u^3 + 27v^2 = 0$ in two points J, K. In J appears a new minimum, an actor. In K the newly appeared actor captures the old one, K is the catastrophe point for capture. But if we continue to describe the unit circle (C), we see that, after a turn, the predator - in a hungry state - becomes its prey! This apparently paradoxical statement may in fact involve the explanation of a considerable amount of facts in Mythology (The Werwolf), in Ethnology (Hunting rituals involve in general simulation of the prey by the hunters), in magic thinking in general. ... '

From: A global dynamical scheme for vertebrate embryology.

By René Thom

Further Reading

(Partly based on the bibliography in [24].)

Note: T. T. B. = Towards a Theoretical Biology, ed. C. H. Waddington, Edinburgh (1970-1972).

By René Thom

1. Stabilité Structurelle et Morphogenèse, Benjamin, New York (1972).
2. Une théorie dynamique de la morphogenèse, T. T. B. 1, 152-66.
3. with C. H. Waddington: Correspondence, T. T. B. 1, 166-79.
4. Topologie et signification, l'Age de Science 4 (1968), 219-42.
5. Topological models in biology, T. T. B. 3, 89-116 and Topology 8 (1969), 313-36.
6. A mathematical approach to morphogenesis: archetypal morphologies, Wistar Institute Symposium Monograph No. 9 (1969), 165-74.
7. Topologie et linguistique, Essays on Topology and Related Subjects, Springer-Verlag (1970), 226-48.
8. Structuralism and biology, T. T. B. 4, 68-82.
9. Modèles mathématiques de la morphogenèse, Ch. 1-3, mimeographed, I. H. E. S. (1970/71).
10. Les symétries brisées en physique macroscopique et la mécanique quantique, (to appear).
11. A dynamical scheme for vertebrate embryology, A. A. A. S., Philadelphia (1971).
12. Sur les équations différentielles multiformes et leurs intégrales singulières, (to appear).
13. Sur la typologie des langues naturelles: essai d'identification psycholinguistique, in Formal Analysis of Natural Languages, ed. Menton, Proc. I. R. I. A. Congress (1970), Janua Linguarium (to appear).

14. Language and catastrophes, Proc. Internat. Sympos. in Dynamical Systems (Salvador 1971), ed. M. Peixoto, Academic Press, New York (1973).

15. La théorie des catastrophes: état présent et perspectives, Manifold 14, see [42].

By E. C. Zeeman

16. Geometry of catastrophe, Times Literary Supplement, 10 December, 1971.

17. Differential equations for the heartbeat and nerve impulse, T. T. B. 4, 8-67.

18. A catastrophe machine, T. T. B. 4, 276-82.

19. Catastrophe theory in brain modelling, Conference on Neural Networks, I. C. T. P., Trieste (1972), (to appear in J. Neuroscience).

20. with C. Isnard, Some models from catastrophe theory, Conference on Models in Social Sciences, Edinburgh (1972), (to appear).

21. with P. J. Harrison, Applications of catastrophe theory to macroeconomics, Symposium on Applications of Global Analysis, Utrecht (1973), (to appear).

22. On the unstable behaviour of stock exchanges, mimeographed, Warwick (1973), (to appear in J. of Math. Economics).

23. Applications of catastrophe theory, preprint (1973).

Miscellaneous

24. R. Abraham, Introduction to morphology, Quatrième Rencontre entre Mathématiciens et Physiciens (1972) Vol. 4, Fasc. 1, Publ. du Département de Mathématiques de l'Université de Lyon, Tome 9 (1972), Fasc. suppl. 1, 38-114.

25. V. I. Arnol'd, On braids of algebraic functions and the cohomology of 'swallowtails', Uspehi Mat. Nauk 23, 4 (1968), 247-8.

26. V. I. Arnol'd, On matrices depending on parameters, Russian Math. Surveys 26, 2 (1971), 29-43, (translated from Uspehi Mat. Nauk 26, 2 (1971), 101-14).

27. V. I. Arnol'd, Lectures on bifurcations in versal families, Russian Math. Surveys 27, 5 (1972), 54-123, (translated from Uspehi Mat. Nauk 27, 5 (1972), 119-84).

28. V. I. Arnol'd, Integrals of rapidly oscillating functions and singularities of projections of Lagrangian manifolds, Functional Anal. Appl. 6 (1973), 222-4, (translated from Funkcional. Anal. i Priložen. 6, 3 (1972), 61-2).

29. V. I. Arnol'd, Normal forms for functions near degenerate critical points, the Weyl Groups of A_k, D_k, E_k and Lagrangian singularities, Functional Anal. Appl. 6 (1973), 254-72, (translated from Funkcional. Anal. i Priložen. 6, 4 (1972), 3-25).

30. V. I. Arnol'd, Classification of unimodal critical points of functions, Funkcional. Anal. i Priložen. 7, 3 (1973), 75-6.

31. N. A. Baas, On the models of Thom in biology and morphogenesis, lecture notes, Virginia (1972).

32. J. Bochnak and T.-C. Kuo, Different realizations of a non sufficient jet, Indag. Math. 34 (1972), 24-31.

33. D. H. Fowler, The Riemann-Hugoniot catastrophe and van der Waals equation, T. T. B. 4, 1-7.

34. D. H. Fowler, Translation of [1] into English, (to appear).

35. A. M. Gabrielov, Intersection matrices for certain singularities, Funkcional. Anal. i Priložen. 7, 3 (1973), 18-32.

36. A. N. Godwin, Three dimensional pictures for Thom's parabolic umbilic, Inst. Hautes Etudes Sci. Publ. Math. 40 (1971), 117-38.

37. J. Guckenheimer, Bifurcation and catastrophe, Proc. Internat. Sympos. in Dynamical Systems (Salvador 1971), ed. M. Peixoto, Academic Press, New York (1973).

38. J. Guckenheimer, Catastrophes and partial differential equations, Ann. Inst. Fourier 23 (1973), 31-59.

39. J. Guckenheimer, Review of [1], Bull. Amer. Math. Soc. 79 (1973), 878-90.

40. T.-C. Kuo, On C^0-sufficiency of jets of potential functions, Topology 8 (1969), 167-71.

41. K. Jänich, Caustics and catastrophes, Math. Ann. 209 (1974), 161-80.

42. Manifold 14, Spring 1973, (Manifold Publications, Math. Inst., Univ. Warwick, Coventry, England).

43. J. Mather, Right equivalence, manuscript.

44. F. Pham, Remarque sur l'equisingularité universelle, mimeographed, Université de Nice (1970).

45. F. Pham, Classification des singularités, Douzième Rencontre entre Mathématiciens et Physiciens (1971) R. C. P. no. 25, Strasbourg.

46. I. R. Porteous, The normal singularities of a submanifold, J. Differential Geometry 5 (1971), 543-64.

47. T. Poston and A. E. R. Woodcock, On Zeeman's catastrophe machine, Proc. Camb. Phil. Soc. 74 (1973), 217-226.

48. T. Poston and A. E. R. Woodcock, A geometrical study of the elementary catastrophes, Lecture Notes in Mathematics No. 373, Springer-Verlag (1974).

49. L. S. Shulman and M. Revzen, Phase transitions as catastrophes, Collective Phenomena 1 (1972), 43-7.

50. D. Siersma, Singularities of C^∞ functions of right-codimension smaller or equal than eight, Indag. Math. 25 (1973), 31-7.

51. F. Takens, A note on sufficiency of jets, Invent. Math. 13 (1971), 225-31.

52. F. Takens, Singularities of functions and vectorfields, Nieuw Arch. Wisk. (3), XX (1972), 107-30.

53. F. Takens, Unfoldings of certain singularities of vector fields, generalized Hopf bifurcations, J. Differential Equations 14 (1973), 476-93.

54. G. N. Tyurina, Resolution of singularities of plane deformations of double rational points, Functional Anal. Appl. 4 (1970), 68-73, (translated from Funkcional. Anal. i Priložen. 4, 1 (1970), 77-83).

55. G. Wassermann, Stability of unfoldings, Dissertation, Regensburg, Lecture Notes in Mathematics No. 393, Springer-Verlag (1974).

56. G. Wassermann, (r, s)-stability of unfoldings, preprint.

List of symbols

$\tilde{A}$	Germ of the set A 1
$\alpha!$	$\alpha_1!.\alpha_2!\ldots\alpha_n!$ where $\alpha = (\alpha_1, \alpha_2, \ldots, \alpha_n)$ 29
$\lvert\alpha\rvert$	$\alpha_1 + \alpha_2 + \ldots + \alpha_n$, α as above 29
$\tilde{f}$	Germ of f 1
$\hat{f}$	Jet of f 31, 34
f*	Homomorphism induced by f 37
K(V)	Quotient field $Q(K[x]/\mathfrak{n}(V))$ 107
K[x]	Ring of polynomials in $x_1, \ldots, x_n$ over K 104
$\Sigma^{i_1,\ldots,i_n}$	Boardman-Thom singularity set 78
$(x, y)^{\alpha,\beta}$	$x_1^{\alpha_1} \ldots y_k^{\beta_k}$ 30
$\langle\sim\rangle_{\mathcal{R}}$	Ideal generated by ~ over $\mathcal{R}$ 31
$\langle\partial f/\partial x_i\rangle$	$\langle\partial f/\partial x_1, \ldots, \partial f/\partial x_n\rangle_{\mathcal{E}(n)}$ 94
$\langle\partial\eta\rangle$	$\langle\partial\eta/\partial x_i\rangle$ 124
$(r, \tilde{f})$	r-parameter unfolding $\tilde{f}$ 120
$\{..\}^{\sim}$	Germ of {..} 1, 101
√	End of proof 5

Index